Proceedings of the U.S. National Workshop on
STRUCTURAL CONTROL RESEARCH

Edited by

G.W. Housner and S.F. Masri

October 1990

Proceedings of the

U.S. National Workshop on

STRUCTURAL CONTROL RESEARCH

25 - 26 October 1990
University of Southern California
Los Angeles, California

Convened by

California Universities for Research in Earthquake Engineering

Sponsored by

U.S. NATIONAL SCIENCE FOUNDATION

Edited by

George W. Housner
Div. of Engin. & Applied Science
California Institute of Technology
Pasadena, California 91125

Sami F. Masri
Dept. of Civil Engineering
University of Southern California
Los Angeles, California 90089

October 1990

(USC Publication No. CE-9013)

Library of Congress Cataloging–in–Publication Data

U.S. National Workshop on Structural Control Research (1990 :
University of Southern California)
  Proceedings of the U.S. National Workshop on Structural Control
Research : 25-26 October 1990, University of Southern California,
Los Angeles, California / edited by George W. Housner, Sami F.
Masri.
    p.   cm.
"October 1990."
Sponsored by U.S. National Science Foundation.
"USC publication no. CE–9013."
ISBN 0-9628908–0–4
1. Structural control (Engineering)- -Congresses. 2. Intelligent
control systems - -Congresses. I. Housner, G. W. (George William),
1910–   . II. Masri, S. F. III. National Science Foundation (U.S.)
IV. Title.
TA654.9.U2   1990   91–2839
624. 1'7- -dc20   CIP

ISBN 0-9628908-0-4

USC Publication No. CE-9013

COPYRIGHT © December 1990
DEPARTMENT OF CIVIL ENGINEERING
SCHOOL OF ENGINEERING
UNIVERSITY OF SOUTHERN CALIFORNIA
LOS ANGELES, CALIFORNIA 90089-2531, U.S.A.

Printed in the United States of America

FOREWORD

The actions of earthquakes and winds, and even man-made and environmental disturbances, can cause unacceptable motions that are objectionable to the occupants of a building or may even be damaging to the structure and its contents. The amplitude of the motions can be reduced by isolating the structure from the sources of excitation, or by introducing counterforces to oppose the motions, or by a combination of the two methods. The method of isolating is called passive control, the method of opposing forces is called active control, and the combination of the two is called hybrid control. The potential benefit of control of structures in seismic regions, of buildings in windy regions, of space structures, etc., dictates that a strong research program should be undertaken. To explore the needs and future directions of such a research program, a workshop was held in October 1990.

ACKNOWLEDGMENTS

The U.S. National Workshop on Structural Control Research was held at the University of Southern California on 25 and 26 October 1990. Approximately one hundred participants with a variety of technical backgrounds participated in the formal presentations and discussions covering current practices, research, and future needs. These Proceedings present the technical papers and summary statements of the formal discussions.

The Workshop expenses were supported by the National Science Foundation through Grant BCS-9001256 to California Universities for Earthquake Engineering Research (CUREe). Additional support was furnished by the National Center for Earthquake Engineering Research (State University of New York at Buffalo), and by the Department of Civil Engineering at the University of Southern California.

The interest and encouragement of Dr. Shih-Chi Liu of the National Science Foundation is much appreciated.

We are indebted to all participants whose attendance and contributions helped us focus on the important and critical issues of structural control research, and whose enthusiasm for acting on the recommendations of the Workshop will provide the motivation to complete the tasks that lie ahead of us. Special thanks are due the Working Groups recorders.

The assistance of Mrs. Janine Nghiem and Mr. Raimondo Betti, of the USC Department of Civil Engineering, in the various phases of the Workshop is greatly appreciated.

The opinions, findings, conclusions and recommendations expressed in these Proceedings are those of the individual contributors and do not necessarily reflect the views of the NSF or other government agencies.

George W. Housner   Sami F. Masri
Caltech   USC

Proceedings of the U.S. National Workshop on

STRUCTURAL CONTROL RESEARCH

Table of Contents

Group # 1: Analytical Methods
Group # 2: Experimental Approaches
Group # 3: Building Applications
Group # 4: Non-Building Applications
Group # 5: Interdisciplinary Approaches
Group # 6: International Coordination
Group # 7: Information Dissemination

Invited Papers

Submitted Papers

Appendices

Appendix A: Workshop Announcement

Appendix B: List of Participants

x

U.S. NATIONAL WORKSHOP
ON
STRUCTURAL CONTROL RESEARCH

---

25 - 26 October 1990
Davidson Conference Center
University of Southern California
Los Angeles, California

Co-Chairmen: G.W. Housner and W.J. Hall

Thursday, 25 October 1990   (Davidson Center)
8:00 a.m.   Registration

8:30 a.m.   (Plenary Session) Welcome and Introduction
             Introduction (G.W. Housner)
             Welcoming Remarks (USC/M.S. Agbabian)
             Welcoming Remarks (USC/Dean Leonard Silverman)
             Welcoming Remarks (NSF/S.C. Liu)
             Purpose of Workshop (G.W. Housner)

9:15 a.m.   Invited Paper #1 (Japan; T. Kobori) (40 min)
             "State-of-the-Art of Structural Control Research in Japan"

10:00 a.m.   Status Reports on Ongoing Research Activities by Leading
             Japanese Companies; (12 min. each)
             (1) [Kajima]/Kobori-Koshika:"Seismic-Response-Controlled
                 Structure with Active Mass Driver Systems"
             (2) [Obayashi]/Suzuki et al: "Active Vibration Control
                 for High-Rise Buildings Using Dynamic Vibration
                 Absorber Driven by Servo Motor"
             (3) [Shimizu]/Ohsaki-Mita:"Current Research Activites of
                 Shimizu for Response Control"
             (4) [Taisei]/Kitamura-Kawamura-Yamada-Fuji:"Structural
                 Response Control Technologies of Taisei Corporation"
             (5) [Takenaka,Kayaba]/Aizawa-Hayamizu-Higashino-Soga-Yamamoto
                 -Haniuda: "Experimental Study of Dual Axis Active
                 Mass Damper"

11:00 a.m.   15 min. coffee break

11:15 a.m.   Invited Paper #2 (Germany; J. Melcher) (40 min)
               "A Survey on German and European Activities in the field of
               Adaptive Structures"

12:00 p.m.   Lunch (USC Town and Gown); Invited Speaker:
               Robert Hanson/NSF; "Remarks on Structural Control"

1:30 p.m.   Invited Paper #3 (USA; T. Soong) (40 min)
               "State-of-the-Art of Structural Control Research in U.S.A."

2:15 p.m.   Status Reports on Ongoing Research Activities;
               (5 - 7 min. each) 15 min. coffee break at 3:30

5:30 p.m.   (Adjourn 1st day meetings)

6:30 p.m.   Social Hour; (USC Town and Gown)

7:00 - 9:00 p.m.   Banquet (USC Town and Gown); Invited Speaker
                   Bruce Murray/[Caltech-JPL]: "Space Stations: How to
                   Make Them Simple"

* *

Friday, 26 October 1990   (Davidson Center)

8:00 a.m. - 10:00 a.m.   Morning Sessions
                         Individual Group Meetings (4 Working Groups)

               Format and agenda for each group:
               1. Opening statement/paper by moderator and/or coordinator
                  (Summary of ongoing work in this area) (15 min)
               2. Discussion of U.S. program in the context of world-wide
                  activities and in relation to IDNDR
               3. Research directions (next five years)
               4. Research agenda and recommended funding levels (next five years)
               5. Concluding remarks

Group 1: **Analytical Methods**   (Davidson CC; Room 1)
Moderator: S. Shinozuka   Coordinator: T. Caughey   Recorder: R. Miller

xii

Group 2: **Experimental Approaches**   (Davidson CC; Room 221A)
   Moderator: B. Wada   Coordinator: J. Kelly   Recorder: D. Foutch

Group 3: **Building Applications**   (Davidson CC; Room 221B)
   Moderator: T. Soong   Coordinator: R. Hanson   Recorder: W.D. Iwan

Group 4: **Non-Building Applications** (Davidson CC; Room 215)
   Moderator: W. Hall   Coordinator: R. Scanlan   Recorder: A. Abdel-Ghaffar

10:00 a.m.   15 min. coffee break

10:15 a.m.   Plenary Meeting   (Chairman: G.W. Housner)
                                                 (Co-Chairman: W.J. Hall)
                      Summary Report of Group #1 (15 min)
                      Summary Report of Group #2 (15 min)
                      Summary Report of Group #3 (15 min)
                      Summary Report of Group #4 (15 min)

11:15 a.m.   General Discussion

12:00 p.m.   Adjourn Workshop

12:15 p.m.   Lunch; (USC Town and Gown); Invited Speaker
                      John Vostrez/[Caltrans]: "Intelligent
                      Transportation Systems"

* *

PURPOSE OF THE WORKSHOP

In recognition of the growing awareness by civil engineers worldwide of the potential of active (hybrid) protective systems for natural hazard mitigation, a "Panel on Structural Control Research" was established to (1) develop a plan for a U.S. program in the active control of civil structures and (2) develop a plan under the auspices of the International Decade for Natural Disaster Reduction for U.S. participation in collaborative international research in the active control field. Further details concerning the Panel are given in a separate section of these Proceedings.

In order to carry out its responsibilities, the U.S. Panel convened a Workshop on Structural Control Research with the following major objectives:

1. To summarize the state of the art of the structural control field.

2. To identify and prioritize needed research in the field.

3. To develop preliminary plans for the analytical and experimental advancement of the field and for the performance of full-scale testing.

4. To facilitate the transmission of information concerning state-of-the-art developments in this rapidly evolving field.

The Workshop program was designed for researchers, engineers, architects, and others who are interested in the theory, experimentation, and application of the vibrational control of structures, sensitive equipment, etc., under the action of dynamic loads such as earthquake and wind.

The first day of this two-day Workshop featured invited papers and summary reports by U.S. and international representatives from universities, research organizations and industrial concerns. Presentations about ongoing research activities or future research plans in the general area of adaptive structural systems were solicited. Among the specific topics addressed in the Workshop were: possible applications, reliability issues, combined active/passive approaches, large scale tests, economic issues, retrofitting problems, implementation techniques, control purpose (safety/comfort), dynamic environment (wind, earthquake, etc.), actuator development, distributed sensor technology, control energy sources, control algorithms, active parameter control methods, stability considerations, system identification

1

procedures, computational issues, architectural considerations, educational/training issues, and international collaborative projects.

The second day of the Workshop consisted of several workshop panels that discussed topics related to specific areas such as (1) analytical methods, (2) experimental approaches, (3) building applications, and (4) non-building applications. The Workshop culminated in a prioritized research plan by each of the four Workshop Working Groups, which are included in these Proceedings.

The University of Southern California undertook the task of convening the workshop and publishing the proceedings. The guidance as well as most of the funding have come from the National Science Foundation through the Earthquake Engineering Program.

George W. Housner   Sami F. Masri
Caltech   USC

U.S. PANEL ON STRUCTURAL CONTROL RESEARCH

by

George W. Housner
Div. of Engin. & Applied Science
California Institute of Technology
Pasadena, California 91125

Sami F. Masri
Department of Civil Engineering
University of Southern California
Los Angeles, California 90089-2531

ABSTRACT

An overview is presented of the recent establishment of a panel in the United States under the auspices of the National Science Foundation to deal with the field of active structural control. Information is given about the purpose of the panel, its organizational structure, future plans, and some proposed international collaborative efforts.

INTRODUCTION

In recognition of the growing worldwide awareness by civil engineers of the potential of active protective systems for earthquake hazard mitigation, the recent 1988 World Conference on Earthquake Engineering in Tokyo-Kyoto convened a one-day Special Theme Session on "Seismic Response Control of Structural Systems." These control systems have the potential for significantly reducing the dynamic forces experienced by structures under arbitrary excitations.

Key items of discussion during the aforementioned meeting, which had enthusiastic approval from the conference participants, were the need for organized national programs and the need for cooperative multi-national efforts involving collaborative projects, exchange of personnel, jointly organized tests, and exchange of data and technical information.

BACKGROUND INFORMATION

With the above discussion in mind, the Editors of this "Proceedings" prepared a proposal for CUREe to establish a U.S. panel on active structural control which was submitted to the Earthquake Hazards Mitigation Program of the National Science Foundation. The Panel activities were funded through the Structural Systems Program, Dr. S.C. Liu Program Manager.

3

The managing organization of the Panel activities is CUREe (California Universities for Research in Earthquake Engineering) which was formed in 1988 as a not-for-profit corporation. At present, the Institutional Members of CUREe are: Caltech, UC-Berkeley, UC-Davis, UC-Irvine, UCLA, UC-San Diego, Stanford, and USC.

PURPOSE OF THE PANEL

This panel is charged with the following responsibilities:

1. To develop a plan for a U.S. program in the active (hybrid) control of structures.

2. To conduct, in collaboration with a counterpart panel in Japan established by the Japan Science Council, a joint U.S.-Japan cooperative work plan for research, development, and implementation of controlled structural systems.

3. To develop a plan of action under the auspices of the International Decade for Natural Disaster Reduction for U.S. participation in collaborative international research in active control.

4. To monitor U.S. efforts in this area.

In order to carry out its responsibilities, the U.S. Panel first convened this U.S. National Workshop on Structural Control Research. The objectives of this Workshop are given a separate section of the Proceedings.

RECOMMENDATIONS

The Working Groups provided lists of research topics and other activities that were considered important. These are discussed in the reports that follow. It is recommended that these topics form the basis of a program in active control of structures in the United States.

In order to avoid falling hopelessly behind in research in the important area of structural control, the U.S. should allocate at least as much funding in this area as does Japan. Funding at a level of $5 million per year is recommended to cover all phases of the structural control program, which includes analytical methods, experimental approaches, building applications, non-building applications, interdisciplinary approaches, international coordination, and information dissemination.

ORGANIZATION OF THE PANEL

The panel is composed of seven members whose names and affiliations are listed below:

1. Mr. Frank Conati, MTS Systems Corporation, Minneapolis, Minnesota

2. Professor William J. Hall, University of Illinois, Urbana, Illinois

3. Professor George W. Housner, California Institute of Technology, Pasadena, California

4. Professor Sami F. Masri, University of Southern California, Los Angeles, California

5. Professor Masanobu Shinozuka, Princeton University, Princeton, New Jersey

6. Professor Tsu T. Soong, State University of New York at Buffalo, Buffalo, New York

7. Mr. Ben K. Wada, Jet Propulsion Laboratory, Pasadena, California

An Executive Committee consisting of G.W. Housner, S.F. Masri, and T.T. Soong oversees the Panel activities. The objectives of the Panel will be accomplished through the efforts of seven Working Groups:

1. Analytical Methods

2. Experimental Methods

3. Building Applications

4. Non-Building Applications

5. Interdisciplinary Approaches

6. International Coordination

7. Information Dissemination

5

Figure 1 shows an overall organizational chart for the Panel. The membership list for the Executive Committees of the seven Working Groups is given in Table 1.

COORDINATION WITH JAPAN

Researchers in Japan are seriously studying the active control of structures. The Japan researchers have indicated that they would like to coordinate the Japan efforts with those in the U.S. They have organized a committee on Active Control of Structures in the Japan Science Council and plan to establish a Workshop Panel to operate in parallel with the U.S. Panel.

Several of the Japan workers attended the U.S. Workshop and it is planned to send several of the U.S. workers to attend the first Japanese Workshop. In addition, it is planned to maintain liaison by visits to Japan in each of the following two years of the program. A tentative diagram indicating a proposed organizational chart for U.S.-Japan collaborative efforts in the field of structural control research is shown in Figure 2.

PANEL REPORT

Based on the results of this U.S. National Workshop and a similar workshop that the counterpart Japan Panel is planning to convene, the U.S. Panel will prepare a report containing long-range cooperative plans for a U.S. program in structural control research, development, and implementation.

INTERNATIONAL COLLABORATION

In keeping with the spirit and intent behind its creation, the U.S. Panel would like to pursue with all interested countries possible cooperative multi-national efforts involving conferences and workshops, collaborative projects and the exchange of technical information.

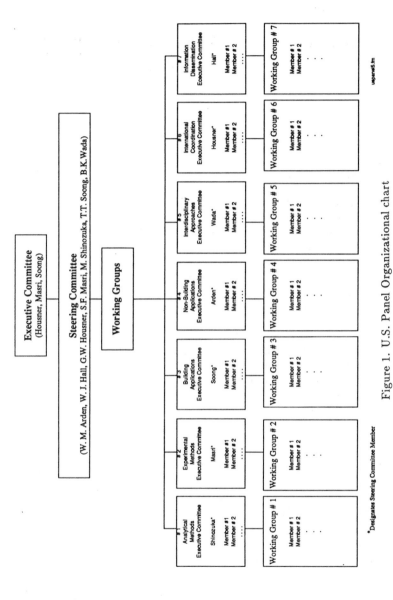

Figure 1. U.S. Panel Organizational chart

7

Table 1. Membership list of Working Groups Executive Committees

---

U.S. PANEL ON STRUCTURAL CONTROL RESEARCH

Executive Committees of Working Groups

Note: (.)* Designates Panel Steering Committee Member

1 Analytical Methods

Professor Masanobu Shinozuka* (Chairman)
Department of Civil Engineering
Princeton University
Princeton, New Jersey 08544

Professor L. Bergman
Department of Aeronautics
University of Illinois
Urbana, Illinois 61801

Professor Thomas K. Caughey
Division of Engineering and Applied Science
California Institute of Technology
Pasadena, California 91125

Professor Billie F. Spencer, Jr.
Department of Civil Engineering
University of Notre Dame
Notre Dame, Indiana 46556

Professor J.N. Yang
Department of Civil Engineering
The George Washington University
Washington, D.C. 20052

2 Experimental Methods

Professor Sami F. Masri* (Chairman)
Department of Civil Engineering
University of Southern California
Los Angeles, CA 90089-2531

Professor Douglas A. Foutch
Department of Civil Engineering
MC-250
University of Illinois
Urbana, Illinois 61801

Mr. Neal R. Petersen
MTS Systems Corporation
Minneapolis, Minnesota 55424

Professor Andrei Reinhorn
Department of Civil Engineering
State University of New York
Buffalo, NY 14260

3 Building Applications

Professor Tsu T. Soong* (Chairman)
Department of Civil Engineering
State University of New York at Buffalo
Buffalo, NY 14260

Professor G.C. Hart
Department of Civil Engineering
University of California
Los Angeles, California 90024

Professor James O. Jirsa
Department of Civil Engineering
The University of Texas
Austin, Texas 78758

Professor J.M. Kelly
Department of Civil Engineering
University of California
Berkeley, California 94720

9

4 Non-Building Applications

Mr. Frank Conati* (Chairman)
MTS Systems Corporation
Minneapolis, Minnesota 55424

Professor Ahmed M. Abdel-Ghaffar
Department of Civil Engineering
University of Southern California
Los Angeles, California 90089-2531

Professor Bruce M. Douglas
Civil Engineering Department
University of Nevada
Reno, NV 89557

Professor Robert H. Scanlan
Department of Civil Engineering
The Johns Hopkins University
Baltimore, Maryland 21218

5 Interdisciplinary Approaches

Mr. Ben Wada* (Chairman)
Jet Propulsion Laboratory
Mail Station 157/507
Pasadena, California 91109

Professor L. Meirovitch
Department of Engineering Science & Mechanics
Virginia Polytechnic Institute and State University
Blacksburg, Virginia 24061-0219

Dr. Zolan Prucz
Modjeski and Masters
1055 St. Charles Ave
New Orleans, Louisiana 70130

10

6 International Coordination

Professor George W. Housner* (Chairman)
Division of Engineering And Applied Science
California Institute of Technology
Mail Code 104-44
Pasadena, California 91125

Professor Robert D. Hanson
Director, Biol. & Critical Systems Div.
National Science Foundation
1800 G Street, Rm 1130
Washington, DC 20550

Professor W.D. Iwan
Division of Engineering and Applied Science
California Institute of Technology
Pasadena, California 91125

7 Information Dissemination

Professor William J. Hall* (Chairman)
Department of Civil Engineering
University of Illinois
Urbana, Illinois 61801

Professor Richard K. Miller
Department of Civil Engineering
University of Southern California
Los Angeles, California 90089-2531

Professor James T.P. Yao
Department of Civil Engineering
Texas A & M University
College Station, Texas 77843-3136

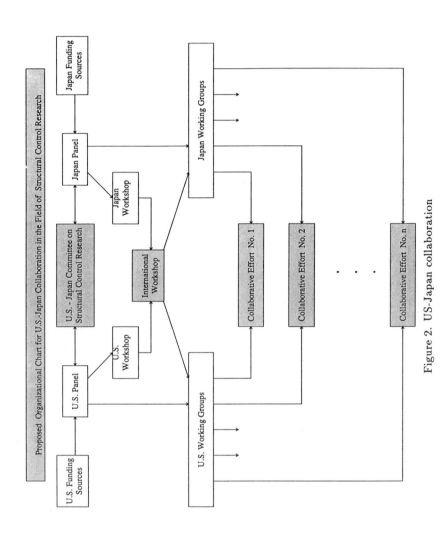

Figure 2. US-Japan collaboration

12

U.S. NATIONAL WORKSHOP ON
STRUCTURAL CONTROL RESEARCH

REPORT OF WORKING GROUP # 1

ANALYTICAL METHODS

Moderator: M. Shinozuka
Coordinator: T.K. Caughey
Recorder: R.K. Miller

Introduction

The majority of the Workshop participants elected to attend this Working Group. Consequently, a small panel was created to initiate the deliberations. The panel members were: L. Bergman, T.K. Caughey, R.K. Miller, B. Spencer, and J.N. Yang. The session was opened by the moderator who began by offering a statement of the basic problem in structural control, paraphrased here as follows:

"To add control devices to structures to limit the response during service and extreme loading conditions, in order to (1) prevent failure, (2) minimize damage, (3) protect secondary systems, and (4) provide comfort for building occupants." The central role of appropriate cost/benefit analyses in structural control was emphasized.

The moderator then presented two viewgraphs with a suggested outline for the discussion of technical issues related to structural control. These viewgraphs, the contents of which are reproduced below, divide the research topics into three basic areas: (1) passive control systems, (2) active control systems, and (3) hybrid control systems.

View Graph Summary

- Overall Technical Issues

    - Integrated Design of Structures and Control Systems

    - Reality Checks: Practicality of Sensing, Computing, and Actuator Requirements

13

- Passive Control Systems

  - System Modeling

    * Micro-modeling (components and devices)
    * Macro-modeling (complete structural systems)

  - System Identification

    * initial or virgin-state system
    * in-service or partially deteriorated system

  - Uncertainty and Reliability

  - Innovative Systems (other than base isolation)

  - Design Code or Design Criteria

    * verification
    * certification

  - Cost Effectiveness

    * Safe-life
    * Damage (secondary systems)
    * Serviceability

- Active Control Systems

  - Modeling

    * Structural Systems
      · Discrete models
      · Continuous models
    * Actuator Dynamics
    * Variable Damping and Stiffness
    * Others (sensors, etc.)

  - System Identification

    * System degradation
    * Damage identification

  - Control Theory and Algorithms

    * Linear and Nonlinear Structures

14

* Time Domain and Frequency Domain Techniques
    · Time Delay
    · System Uncertainty
    · Loading Uncertainty
    · Soil-Structure Interaction Effects
    · Robust Control
  − Sensors and Actuators
    * Number and Location
    * Modal Spill-over Effects
  − Reliability
  − Cost Effectiveness

• Hybrid Control Systems

This outline was addressed by the participants in the discussion. A summary of the relevant issues raised is provided below.

Working Group Panel Members:

The comments of the Working Group Panel members are summarized below.

• Recent developments in *nonlinear control* theory should receive special attention.

• Interaction effects between *actuator dynamics* and structural dynamics should also be explored. Experience within the aerospace engineering field indicates that overlap between the bandwidth of actuators and structures leads to spillover problems.

• There is an acute need in civil engineering applications to focus on control schemes which are *robust* in the sense that they are tolerant of uncertainties in the model of the structure, loading, etc.

• Active mass dampers may not be well suited to problems of seismic protection due, for instance, to excessive power requirements. In such cases, the more beneficial approach may be to attempt to limit energy input to the building.

15

- *Redundancy in sensors* is a requirement of major importance in civil engineering applications.

- Recent developments in *"smart structures"* research, including active elements embedded within truss elements, should be given more attention within the civil engineering community.

- Care must be exercised in testing to characterize uncertainty in a statistical sense in order for *probabilistic models* to be constructed for analysis.

- Not all time delay effects originate with actuator dynamics; for example, a *pure time delay* is introduced due to computational time needed to determine actuator signals appropriate for the control algorithm in use.

- Significant unexplored potential exists for applications of *adaptive control algorithms.*

- Considerable *similarity* exists *between* fundamental problems in *system identification and* in *design* of passive systems. More effort to capitalize on this may lead to novel passive designs to supplement current focused efforts in base isolation.

- Significant benefits in performance may be obtained by fully *integrating design of structures and control systems* into a seamless process.

Open Discussion:

A summary of the open discussion is listed below:

1. **Passive Control Devices**

   - Passive control devices should be incorporated into the original design of the structure, and should not be considered as "add ons" except in special cases, such as retrofits.

   - Passive energy dissipators are most promising for future applications.

   - Guidelines for the design and integration of passive devices into existing structures are badly needed.

   - In some regions, such as Mexico City, base isolation is not a good idea, and other passive devices are of particular importance.

16

- Interstory shear actuators are a promising class of devices for further study.

- More research on dissipative devices, such as those based on thermally-induced metal plasticity, may prove particularly fruitful.

- Strain energy may be used to identify optimal locations for energy dissipation devices within a structure.

- Emphasis should be placed on non-destructive testing of structures to determine accurate and realistic mathematical models of structural behavior.

- Modeling and system identification must be considered as a coupled task, and may involve decisions regarding the use of finite element, lumped parameter, modal, and continuous model descriptions.

- Model verification must be given great importance, since in reality "you don't control the model, you control the structure."

- Parameter and model uncertainty are of particular importance, and must be quantified whenever testing is performed.

- Control algorithms which are robust against uncertainty and variation in model parameters must be developed.

- Future research should focus in part on the development of a realistic representation of structural behavior, taking into account the degree of deterioration, and recognizing the continuous nature of the structure, for both initial (virgin-state) and in-service conditions.

2. Active Control Systems

- Actuator dynamics, including modeling and identification, is an area in need of further research.

- Performance robustness vs. stability robustness: further research on the trade-offs inherent in improving performance and improving stability for large civil structures is needed.

- While there exist robust control algorithms involving co-located sensors and actuators, more research is needed to develop algorithms which improve performance over that obtainable with co-located approaches without sacrificing stability, particularly for nonlinear hysteretic structures.

- The research community would benefit from a broader education for civil engineers: many useful control approaches exist in related disciplines, such as electrical engineering, aerospace engineering, mechanical engineering, and acoustics which are not yet familiar within the civil engineering community.

- Recent developments in nonlinear control theory may permit substantial progress in developing efficient control algorithms for hysteretic structures.

- While chaos is not likely to be of primary importance in practical problems, the quantitative use of statistical models of chaos may provide some interesting new results in structural control.

- Soil-structure interaction effects are important in that accurate definitions base motions are necessary for many approaches within structural control.

- The development of accurate, reliable and inexpensive displacement sensors would provide a substantial benefit to many approaches in structural control.

- The torsional response of buildings is an area of continuing research need.

3. Hybrid Control Approaches

- Hybrid control approaches seem to offer opportunities for improving performance over either active or passive approaches taken individually.

- While passive (base isolation) systems seem to provide advantages for earthquake protection, and active devices (tuned-mass dampers, etc.) seem to offer advantages in controlling wind-induced vibrations, hybrid systems would seem an attractive approach for a wide range of excitations.

- Hybrid control approaches can lead to designs which reduce required actuator forces.

- Friction effects in actuators are always present, and are not well modeled or understood.

18

- More effort should be applied to develop solutions for retrofit problems since the largest need for structural control and hazard mitigation will likely emerge for this class of problems.

4. Recommended Funding Levels

Much analytical research is needed in the early stages of the program. This will provide a basis for planning experimental research and building applications. A budget of $1.5 million per year is recommended for the first three years of the program.

U.S. NATIONAL WORKSHOP ON
STRUCTURAL CONTROL RESEARCH

REPORT OF WORKING GROUP # 2

EXPERIMENTAL APPROACHES

Moderator: B. Wada
Coordinator: J. Kelly
Recorder: D. Foutch

Opening Discussion

Active control of structures under dynamic loads involves four elements: the sensors, the controller, the actuator system and the structure. Each of these elements contain uncertainties and performance characteristics that are less than ideal. As a consequence, it is not possible to separate the analytical study of a method to attenuate adverse response from the imperfect nature of the components if the study is to be effective. Therefore, experimental investigations must go on simultaneously with analytical ones.

Testing is needed in order to understand the realistic interaction of the four elements (sensors, controller, actuator and structures) and to evaluate their abilities to meet the control objectives. Experiments are needed to validate control theories. Experimental research also opens new avenues of research. The experience of researchers at SUNY Buffalo was that each time they moved to a larger scale, new unknowns and complexities were revealed that required additional research.

Experimental research using small-, medium- and full-scale models is important. Small-scale tests are relatively inexpensive and are useful in the initial stages of developing a control system. They contain all of the physical elements involved in the prototype but they are much less expensive to conduct. Therefore, many control systems may be tested for a relatively small amount of money. Systems that are clearly not feasible or practical can be discarded. As the scale of the structure becomes larger, the physical components of the control system become more realistic and more useful tests can be conducted. The final tests of any control system should

20

be done at full-scale on a realistic structure.

Experimental Research in the United States

A great amount of experimental research on active control has been done in the U.S. However, very little of this has been devoted to civil engineering structures. This is an important issue because the uncertainties associated with each element of the control system are much greater for civil engineering systems than they are for electrical or aerospace systems. Thus, much of the research in these areas is not directly transferable to our problems.

A complete list of experimental research done in the U.S. is not available since all institutions conducting such research were not represented at the workshop.

Researches at the University of Southern California have tested two different active control systems on very small-scale models. Researchers at the University of California at Berkeley have tested schemes involving active control of base-isolated buildings using small-scale models. Several active control systems have been tested by researchers at SUNY Buffalo using models ranging in size from small-scale to full-scale. The full-scale tests were done in Japan in cooperation with Japanese researchers. Fundamental issues in active control were investigated by researchers at the University of Illinois at Urbana-Champaign. Descriptions of these activities may be found in these proceedings.

Research on passive control systems has also been conducted in the U.S. This work has been conducted at UC Berkeley, SUNY Buffalo and the University of Michigan. Research on passive systems for retrofit of nonductile RC frames is currently underway at the University of Illinois at Urbana-Champaign. Passive systems have been installed in several tall buildings in the U.S. to reduce response to wind loads.

Research on active and passive control in Japan is more extensive than in the U.S. Several Japanese construction companies have developed systems that have been tested on both small-scale and large-scale structures. Some of these involve hybrid systems composed of both active and passive elements. At least one building in Japan has been equipped with an active control system to reduce its response to wind loading. Several full-scale test structures have been built with active and/or hybrid systems to resist earthquake loadings.

21

The European research community has been very active in developing advanced sensors that may be used as part of a control system. Some of these sensors utilize new materials that have been developed for that purpose.

Research Directions

Since relatively little experimental research has been done in active or passive control systems, a great deal of effort needs to be devoted to this activity. Only a relatively small amount of time was devoted to group discussions, so some important research needs were probably overlooked and will go unmentioned in this report. The items mentioned here are not necessarily listed in order of importance.

Component Tests - Standardized tests for each component of an active control system need to be developed. The performance of different pumps, actuators and servo-hydraulic systems need to be measured. This would include the determination of the force-displacement-time relationships of various combinations of these elements in a control system. The performance of various sensors and computer control systems should also be evaluated. The errors and time-delay characteristics associated with each element of the control system need to be quantified. The reliability of these elements needs to be ascertained. A centralized distribution center should be established so that the results of the component tests can be disseminated as quickly as possible for incorporation into the theoretical studies.

Component Development - New sensors need to be developed and sensors used in other applications need to be tested for their suitability for use in active control system. Basic questions on what structural responses (accelerations, velocities, displacements, member forces) can be measured and with what accuracy need to be answered. Development of more compact and efficient actuators may also be useful. Hardware development in conjunction with control algorithm development should proceed simultaneously. The use of single component, cascaded component and parallel systems may be useful.

Model Testing - All control systems need to be verified in the laboratory. A number of facilities exist in the U.S. where control theories can be tested on small-scale structures. Investigators doing purely theoretical research should be encouraged to affiliate with researchers with access to experimental facilities where their theories may be tested. Joint proposals may be desirable. Any control sys-

22

tem should be tested using mathematical models which realistically represent the expected deviation of the response of the full-scale structure at high levels of excitation.

Promising control systems should be tested on medium-scale structures. Three or four facilities exist in the US with earthquake simulators large enough to test 1/4- to 1/3-scale buildings. These tests should include multi-directional input motions and perhaps structural or nonstructural elements that will behave in a nonlinear manner. Perhaps one or two laboratories could build or refurbish existing medium-scale models that could then be made available to any researcher who wishes to verify a control system. This would avoid the necessity of having to build several expensive models. The input forcing functions for which a control system should be tested must range from those whose characteristics match measured data to those which may occur in the future.

Full-Scale Tests - The most promising control systems need to be tested on a full-scale structure. Such a facility may be relatively costly to build and operate, so it should be done in such a way that it can be made available to any researcher who wants to use it. Another possibility would be to encourage cooperative efforts with companies in Japan who have already built full-scale test facilities. Wherever these tests are done, inexpensive and effective methods need to be developed for exciting these structures for testing purposes.

Development of Active Structures - Most control systems are add-ons. Can we integrate existing structural or nonstructural systems of a building into the control system? These could be used to change the stiffness of the building or to distribute the kinetic energy into higher modes of response. Examples might be active braces that may also change in stiffness. It may also be possible to convert existing elements into energy dissipating passive devices. For high rise structures designed to survive high excitations, this approach seems promising and offers new research areas.

Development of Hybrid Systems - Systems which incorporate both active and passive devices may be promising. New passive devices need to be developed for this application, and further testing and development of existing or proposed devices need to be done.

Adaptive Systems - The response of buildings and bridges is inherently nonlin-

23

ear even at relatively small response levels. Therefore, the development of adaptive control systems with experimental verification is encouraged.

Research Agenda

Given that relatively little experimental research has been done to date compared to analytical research, and given the reliance of any active control system on the known and reliable performance characteristics of various mechanical electrical components, it is recommended that experimental research be given a high priority in the research program. Experimental research can be very expensive, costing perhaps $150,000 to $200,000 per investigator per year, plus equipment costs and the cost of building models. Therefore, it is recommended that some system be developed whereby facilities, equipment and models can be shared by several investigators. One approach is to develop several national testbeds. After the response to a "request for proposals" has been received, ten to twenty of the best proposals should be funded at a level which allows validation using the national testbeds. This would allow for more efficient utilization of scarce funds.

The characterization of standard components of control systems is essential for providing input to theoretical studies. This should be one component of any experimental program. As mentioned above, all promising control systems should be tested in small-scale. This should be done during the first one or two years of the program. The most promising of these (perhaps determined by a panel) should be tested at medium-scale. The hardware and input should possess the uncertainties expected in the prototype situation and all promising techniques to attenuate the response should be tested in the presence of these uncertainties. Finally, one or two of the systems can be tested at full-scale depending on the level of funding available.

Recommended Funding Level

A budget of $1.3 million per year is recommended for the first three years of the program.

24

U.S. NATIONAL WORKSHOP ON
STRUCTURAL CONTROL RESEARCH

REPORT OF WORKING GROUP # 3

BUILDING APPLICATIONS

Moderator: T. Soong
Coordinator: R. Hanson
Recorder: W.D. Iwan

Introduction

The Building Applications working group limited its discussion to the practical aspects of the application of structural control technology to actual building structures. Although many theoretical questions remain to be explored in structural control, the major focus of the working group was on what must be done to translate existing and future theory into practice.

Central Issue

As the discussion developed, it became clear that a central issue which must be resolved is whether structural control systems should be considered life safety systems or merely as a means of improving human comfort and function. Some working group members felt that structural control offered the potential for economical achievement of life safety goals for both new and existing structures. They argued that structural control could, at the very least, offer a "second line of defense" for structural integrity. Other working group members felt that such application was unlikely. They argued that it would not be responsible to remove "tons of steel" and rely instead upon a control device. It was reported that structural control is not currently used for life safety in Japan and that such an application was "very future."

The question of whether structural control should be used for life safety is obviously an issue that must be addressed and resolved before the application of this technology to building structures can be widely recommended or accepted. This

25

is a difficult problem and further information about the effectiveness of structural control will be needed before it can finally be resolved. Nevertheless, the issue must be faced head on and not sidestepped.

Areas Needing Further Research

Setting aside the issue of whether structural control should be relied upon for life safety, the working group turned its attention toward the studies and research needed to bring about the application of control technology to building structures. The major areas which were discussed are listed below. No doubt there are other topics that could be added to the list but these were the topics that surfaced in the limited time allowed for discussion.

Reliability

An important area which requires further study is the reliability of structural control systems, particularly *active* systems. The fundamental questions which need to be answered are what is an acceptable level of reliability for active control systems and what is the attainable level of reliability with current technology. A systems approach to reliability must be adopted which includes not only the control system itself but also the support systems such as power. It was generally believed that structural nonlinear behavior could have a significant effect on reliability. Therefore, reliability studies should incorporate realistic (nonlinear) building behavior.

Nonlinear Structural Behavior

If active control is used for life safety, structural nonlinear behavior must be included in the analysis and design of such systems. The nonlinear models used should be the latest, and most sophisticated, models available. Even when control technology is used only for comfort and function, nonlinear structural behavior may be important. This can be the case for stiff reinforced concrete structures where even relatively small displacements can result in changed structural behavior. More study needs to be devoted to the performance of active control systems including the effects of nonlinear structural behavior.

26

Performance Criteria

Performance criteria for control systems need to be developed before such systems will receive widespread application. These criteria should set out reasonable goals for the performance of different types of control systems under a variety of environmental loads. These criteria need to address passive, active, and hybrid systems. They should also encompass such loads as wind, seismic, and man-made loads. Before control systems can be proposed to building owners, engineers need to know what level of performance can be achieved and at what cost. They also need to know what type of system is best for what application. Considerable research may be required to develop meaningful performance criteria.

Design Criteria and Guidelines

Once performance criteria are established, it will be necessary to develop design criteria and guidelines. This process must involve design professionals as well as university researchers. SEAOC and similar organizations must be able to take ownership of these criteria and standards. Therefore, they must be fully involved in the development process. The process used to develop design criteria for base-isolated structures was presented as a good model for how this objective could be accomplished for structures employing control systems. Any design criteria developed should be based on the latest design and analysis concepts including incorporation of nonlinear behavior. Work in this area needs to begin immediately as it will provide the basis for other needed efforts.

Case Studies

One method for examining such questions as reliability and facilitating the development of performance and design criteria is through case studies. It was recommended that no less than three case study design and analysis projects be undertaken. These case studies should highlight the use of different control technology. They should also encompass existing as well as new buildings and eastern as well as western construction techniques. Such studies would serve to demonstrate just what can and cannot be accomplished with different control system strategies. They would also provide a data base for the development of performance and design criteria. It was felt that three case studies were an absolute minimum.

Scale Model Experiments

More laboratory and field experiments on scale model structures need to be performed in order to gain a better understanding of the practical problems involved in the application of control technology to civil structures. These experimental studies are important to understanding the results of analytical studies and identifying areas requiring further research. One category of structure that clearly requires additional scale model testing is stiff (concrete) structures. So far, control technology has been applied primarily to flexible structures and application to stiff structures does not appear to be well understood. This knowledge gap must to be closed.

Full-Scale Testing

A proposal was presented for an actively controlled demonstration structure at the site of the planned Inventors Hall of Fame in Akron, Ohio. The proposed structure is a slender "theme" tower which would be instrumented with movable masses, movable anchorages, and an active tendon system. The Board of the Hall of Fame has agreed to commit up to $750,000 toward fitting the tower with an active control system if equal matching funds can be obtained from NSF or some other agency.

Working group members supported the concept of a full-scale demonstration project for active control. There was no consensus as to the suitability of the proposed project. It was pointed out that Akron is neither in a highly seismic nor a wind prone region, and the tower is not typical of a building structure. Also, there was an apparent imbalance since the estimated cost of the active control system would be $1.5 million and the reported cost of construction of the tower itself would be only $300,000.

It was agreed that a full-scale demonstration project should be included in the latter phases of the recommended five-year program, and that such a project should have the following characteristics: 1) there must be a set of achievable objectives, 2) the structure selected should resemble a "typical" building structure, and 3) the site should be selected in a region where either wind or earthquakes are a problem.

It was felt that any full-scale demonstration project should be preceded by a formal feasibility study.

A Phased Program

The working group believes that the building applications research and development program should proceed in three phases as indicated in Table 1. Phase 1 would involve analytical and other paper studies of the type referred to in the body of this report. Phase 2 would be devoted to scale model experiments in both the laboratory and field environment. Phase 3 would involve the full-scale demonstration experiment. Generally speaking, these three phases would be carried out sequentially in time. However, there would be some necessary overlap for startup and completion of various elements in the program.

Recommended Funding Level

A budget of $1.0 million per year is recommended for the first three years of the program.

TABLE 1

BUILDING APPLICATIONS

RESEARCH AND DEVELOPMENT PROGRAM

Phase 1	Phase 2	Phase 3
Analytical studies & investigations	Scale model expr.	Full-scale demo project
1. Life safety vs. human comfort 2. Reliability 3. Nonlinear behavior 4. Performance criteria 5. Design criteria/guidelines 6. Case studies Diff. techniques Existing and new strs. (E - W) Stiff or flexible	Lab and field	Clear objectives Typical structure Real loads

U.S. NATIONAL WORKSHOP ON
STRUCTURAL CONTROL RESEARCH

REPORT OF WORKING GROUP # 4

NON-BUILDING APPLICATIONS

Moderator: W. Hall
Coordinator: R. Scanlan
Recorder: A. M. Abdel-Ghaffar

Introduction

Several members of this discussion group suggested that the title, "Non-Building Applications," is inappropriate and should be changed to reflect the purpose of the group to a title such as "Protection of Community" or "Critical Function Facilities." Accordingly, such facilities may be categorized as follows:

1. Structures, other than buildings, which are compact

2. Extended structures

3. Structural contents such as equipment

In any case, the group basically focused on the identification of examples of such non-building applications, types of systems that should be studied and suggestions for needed research and development.

Facilities Appropriate for Structural Control

A considerable amount of time was spent on the identification of non-building structures or critical function facilities. Following is the list of such facilities which are appropriate for structural control:

1. Emergency Response Facilities which include:

   (a) Fire departments

(b) Police departments

(c) Medical and life-support services

(d) Communication facilities serving the above, including emergency-response center activities

(e) Parking structures for the above facilities

2. Transportation Facilities which include:

(a) Bridges, particularly long-span critical bridges

(b) Urban transit services

(c) Railroads and stations

(d) Airports including control towers

(e) Traffic control centers in metropolitan areas

3. Communication services which include:

(a) Telephone facilities which serve both local and long distance areas

(b) Emergency communication facilities

4. Water and sewage facilities which include:

(a) Water purification and distribution plants as well as fire suppression facilities

(b) Sewage treatment plants

5. Banking systems including their computer facilities

6. Power plants (including nuclear power and fossil fuel plants)

7. Industrial facilities including refineries; gas-oil processing and distribution systems; off-shore platforms

8. Harbor and Coastal facilities

9. Gas and liquid fuel facilities

10. Production processing lines critical to public safety

Research Areas

As far as the general research areas are concerned the Group identified three major areas for structural control:

1. Floor systems which may range from

    (a) Small to very large floor systems

    (b) Single to multiple isolation or control systems (active, passive, and hybrid)

2. Special equipment which includes:

    (a) Medical equipment (such as x-ray and nuclear magnetic resonance imaging facilities). Also, power sources in medical facilities which supply life-support and intensive care facilities.

    (b) Computer Systems which range from small to extended facilities and data storage.

    (c) Museum artifacts protection.

    (d) Water plant facilities including cross connection between water and sewage systems.

    (e) Manufacturing equipment.

    (f) Food-supply sources which include warehouse and refrigeration systems.

    (g) Transmission lines including towers and poles

3. Research approaches which include:

    (a) Attainment of mixture of systems (hybrid) including their reliability, cost-effectiveness, and redundancy.

    (b) Development of modern techniques for passive control; also development of new materials for such purposes.

    (c) Adoption of the concept of multi-defense lines.

    (d) Encouragement of new and innovative ideas.

Bridges and Equipment

The following problems have been identified from the various applications to control long-span cable-supported bridges as well as equipment:

1. The optimal design of control actions using optimal control theory.

2. The generation of control action through the use of tendons, mass absorbers, appendages, and actuators or jets.

3. The optimal locations of control actions and sensors on the structures.

4. Proposing several practical control mechanisms for equipment and bridges.

5. The stochastic control of structures considering stochastic dynamic loads of known statistics.

6. Considering the time delay effect in the design of stability analysis of controlled distributed parameters structures subjected to concentrated control actions.

7. The feasibility of using this technique in practical applications. Some of the above problems have been solved and some still need further investigations. Moreover, other problems still need further investigations for the practical implementation of structural control.

The problems are:

1. The identification of the structural parameters of built structures which are needed for designing the control actions and analyzing the controlled response.

2. The combined usage of passive and active (hybrid) control systems to minimize the costs.

3. Devising more control mechanisms efficient for vibration control in particular structures.

4. Control of the nonlinear response in structures.

5. Experimental verification of the analytical results.

33

Proposed Specific Research

1. Study the nonlinearity effects on the response of long-span cable-supported bridges and equipment. Nonlinearities could be geometric, material or due to self-excited forces.

2. Provide means to control the nonlinear response of structures. Compare the use of traditional control methods such as tuned mass dampers and pulse controllers. Investigate the possibility of updating these techniques or devising new ones to provide effective control.

3. Study the relationship between structural identification and structural control considering the uncertainty in estimating the structural parameters and its effect on the effective control. The variations of the structural parameters during control process should also be considered in the design of control actions.

4. The applications of these findings shall be demonstrated on the control of long-span bridges and tall buildings. The plan is to provide a comprehensive control strategy considering time delay effect, nonlinearity effects, and the variations of structural parameters.

Recommendations

Finally, the group recommended that research in the above-mentioned areas be directed toward:

1. Civil agencies (such as the National Science Foundation, the Department of Energy and the Department of Transportation).

2. Military sources; a particular interest is the transfer of relevant technology from military to civilian applications.

3. International cooperation (with public and private agencies) which is essential for exchanging and expanding knowledge.

4. Industry, including construction companies and relevant equipment-manufacturers.

Recommended Funding Level

A budget of $0.6 million per year is recommended for the first three years of the program.

U.S. NATIONAL WORKSHOP ON
STRUCTURAL CONTROL RESEARCH

REPORT OF WORKING GROUP # 5

INTERDISCIPLINARY APPROACHES

Problems similar to the active control of structures during earthquakes and winds are studied in the fields of aerospace, aero-space structures, electrical control theory, noise control, and others. It will be important that researchers in active control of structures should keep informed of similar activities that are underway in these different fields of research and, where feasible, interdisciplinary research projects should be undertaken. A workshop devoted to clarifying the state-of-the-art in these different disciplines should be held.

Recommended Funding Level

A budget of $0.2 million per year is recommended for the first three years of the program.

U.S. NATIONAL WORKSHOP ON
STRUCTURAL CONTROL RESEARCH

REPORT OF WORKING GROUP # 6

INTERNATIONAL COORDINATION

The potential benefit of active control of structures during earthquakes and winds will be applicable to a number of countries, which either face these natural hazards or are actively engaged in engineering structures in hazardous countries. There is much to be gained by international coordination and cooperation. This can be done through international conferences, workshops, or cooperation between committees, panels, or cooperation between individual research workers.

Recommended Funding Level

A budget of $0.3 million per year is recommended for the first three years of the program.

U.S. NATIONAL WORKSHOP ON
STRUCTURAL CONTROL RESEARCH

REPORT OF WORKING GROUP # 7

INFORMATION DISSEMINATION

A program on active control of structures must be supported by appropriate information dissemination both nationally and internationally. This will assist in speeding up the program and providing timely results.

Recommended Funding Level

A budget of $0.1 million per year is recommended for the first three years of the program.

INVITED PAPERS

State-of-the-Art of Seismic Response Control Research in Japan

TAKUJI KOBORI

Professor Emeritus of Kyoto University, Dr. of Eng.
Executive Vice President, Kajima Corporation
6-5-30, Akasaka, Minato-ku, Tokyo 107, Japan

1. Introduction

In conventional earthquake resistant design, most structures are designed to withstand earthquakes of moderate intensity elastically (allowable stress design), and to prevent a collapse during a severe earthquake, thereby preventing loss of human life. However, in many urban areas not only individual buildings, but also entire city functions are becoming intelligence oriented. Therefore, it seems unwise to cling to a design philosophy designated for the severe earthquakes, in which a barely prevented collapse at the structural ultimate limit is recognized as acceptable, providing there is no loss of human life. Is such thinking still acceptable? The conventional philosophy, which is several decades old, can result in a decrease in an individual building's function and loss in its financial value, and prevent re-use of the building after severe earthquakes. This should not be tolerated in the coming age. A technology is required that will not only suppress the vibrations of individual buildings, but will preserve the information and communication function that sustains a city's life.

The seismic response controlled structure is a new concept that is anticipated to fulfill this requirement in the 21st century. With this system, the building itself functions actively and continuously to act on earthquake ground motions. In Japan, there have been many recent technology advances, in this field, and various technologies which were not conceivable a decade ago have now become possible.

This study presents a general view of R&D into active and hybrid control systems relevant to the seismic response controlled structure currently being conducted in the research institutes of private firms and universities in Japan. Reports are presented, in particular, on active and hybrid control devices actually being utilized, focusing on those that have already been installed, and those that are in the experimental stage with the objective of practical application.

2. Conventional Aseismic Design in Japan

2.1 History of earthquake resistant structures in Japan

In Japan, research on earthquake resistant structures began about 100 years ago, in 1891. This was the year the Nobi Earthquake (Magnitude 8.0) occurred. With this earthquake as the impetus, scholars and researchers in the structural engineering field established the basis for the rigid structure concept until the next epochal Great Kanto Earthquake of magnitude 7.9 occurred in 1923. The Nihon Kogyo Bank, whose design was based on the rigid structure theory, suffered no damage during the Great Kanto Earthquake. And also instructions obtained from other many damaged structures contributed to the development of earthquake resistant structures.

After the World War II, the observed strong ground motions were stored with the development of strong motion accelerometers, and owing to the advancement of computers, the characteristics of the dynamic response were able to be analyzed. As a result of these advanced innovations, the 'flexible structure' theory as an expanded concept of the 'rigid' developed, and in 1968, the very first high-rise building in Japan, the 'Kasumigaseki Building', was constructed based on the flexible structure theory. After that many high-rise buildings have been constructed all over Japan.

2.2 Relation between conventional and new seismic design concepts

The relationship among the earthquake resistant, the base isolation and the active seismic response control system is shown in Figure 1.

The fundamentals of the earthquake resistant structure is a rigid structure. This is based on the concept of suppressing as much as possible relative deflection of a structure subjected to large ground motions, by shear wall or bracing. In flexible structural system, as represented by high-rise buildings, the entire structure is constructed flexibly so as to vibrate in higher modes, and the deformations of the structure during earthquakes are dispersed along the height so that relative story deflection between the adjacent floors is suppressed to be small.

The concept of the base isolation system was first introduced in the mid 192Os as an alternative to the earthquake resistant system. With this system, the concept of rigid framing is incorporated to the super-structure, and the vibration is absorbed at the foundation and thus prevented from being transmitted to the upper structure.

Both the flexible structure and the base isolation design concepts consequently utilize the nature of the long natural period of the building to reduce the seismic response.

The earthquake resistant structure and the base isolation system, however, are both endurance and passive types that simply wait for an earthquake to happen, and are unable to take positive action against earthquake motions. In view of the uncertainty of earthquake motions, anxiety regarding safety cannot be dismissed. For example, in the Mexico Earthquake of 1985, great damage was caused by earthquake motions that far exceeded the predictions. These large ground motions were in turn caused by the particular ground conditions. The damage occurred in spite of the revisions to the seismic design criteria (regulation) as shown in Figure 2. Thus, the seismic response control system was conceived with the aim of preventing tragic disasters caused by unpredictable earthquake motions and realizing states of nonstationary and nonresonant of structures. Its ultimate objective is to eliminate severe damage, such as to buildings in high-seismicity zones.

After classification based on design methodology, consequently, rigid structures are designed based on the static method, while other types including the seismic response controlled structures are designed based on the dynamic method in Figure 1.

3. Concept of Seismic Response Control System

3.1 Background

Despite continuous research, the waveforms of earthquake motions are quite unpredictable. The reliability design method, founded on probabilistic theory, is a design method that involves uncertain elements. However, at present it is difficult even to determine a qualitative allowable criteria.

To ensure the safety of buildings subject to highly unpredictable earthquake waveforms, nonlinearity is applied to their structural elements and the path to the nonstationary and nonresonant state is the starting point of the concept of the seismic response controlled structure. Since a definite prediction of an input earthquake waveform is impossible, the only alternative for maintaining safety is to control the building structural part when it receives the input.

In applying this concept, a computer installed in the building performs appropriate recognition and judgment of information transmitted from ground motion sensors and response sensors, and functions to counteract the destructive force of earthquakes. To achieve this, the building selects and changes its own structural dynamic characteristics or induces a controlling force. Thus, damage of the building, including internal information and communication facilities, is prevented even in large earthquakes, and the building is not only maintained in a completely usable state but also offers a pleasant living space in a high residential environment in frequently occurring earthquakes of moderate intensity.

In strong winds, the response of the structure is mainly of low mode, its duration time is long and the control does not need to cope with severe instantaneously changing conditions. Therefore, the function of the vibration control system designed for the earthquake response controlled structure will be sufficiently appropriate for strong wind disturbances.

3.2 Type of seismic response control

A seismic response control system is defined as a system that fortifies the structure by imparting its particular characteristics or the control device that controls the earthquake motions. The procedures are as follows: (1) Cutting off the input energy from the earthquake ground motion, (2) Isolating the natural frequencies of the building from the predominant seismic power components, (3) Providing nonlinear structural characteristics and establishing a nonstationary state nonresonant system, (4) Supplying a control force to suppress the structural response induced by earthquakes, and (5) Utilizing an energy absorption mechanism.

(1) above is the original concept of the ideal goal of base isolation system. If this can be completely realized, procedures from (2) to (5) become unnecessary. However, as this is impossible at present, the currently applied base isolation system is based on (2) with assistance from (5), which offers a longer natural period to the building to evade the resonance. Obviously, the conditions in which this procedure is applied are limited. Next, (2), (3) and (5) are nothing else but the theoretical backbone for realizing high-rise buildings, and it is only possible where the supporting soil base is hard and also where the long period components of the input earthquake ground motions are not predominant.

Seismic response control was originated in how to maintain the building safety against the unpredictable earthquakes without such limited condition. In seismic response control, function of controlling the earthquake ground motions should be given to the structure by adding (5) to (3) or (4).

3.3 Application of seismic response controlled structure

Figure 3 shows examples of applications of the seismic response control system. They can be roughly divided into three categories: passive, active and hybrid.

The base isolation system is positioned under the passive system, and the concept is to isolate the super structure from the foundation. As a vibration system, it depends on the long period of the structure to evade the resonance with a fundamental vibration mode. Therefore, Section 3.2 (2) is applicable, and (5) is supplementary. However, as this system has an natural frequency, needless to say, the scope of the application becomes limited.

Next, the dampers shown in Figure 3, namely, the tuned mass damper, the sloshing liquid damper, the oil damper, the friction damper, the elasto-plastic hysteretic damper and the others function as energy absorbers and are therefore corresponding to Section 3.2 (5). These dampers are also categorized as passive devices, but when the seismic response control of the structure is considered it is practical to incorporate them in the system as much as possible.

The active control system may be of the response control force type or the nonresonant type. The former is based on Section 3.2 (4). In addition to the AMD type, which provides a seismic control force by actively operating the installed mass device, various other methods have been contrived. In these methods, variable tendons are used or a seismic control force is provided by variable damping force which can be generated by the device used in AVS system. All of them are categorized as types that provide a seismic control force. The AVS system is based on Section 3.2 (3) in which a nonstationary nonresonant system is established. It provides nonlinear characteristics to the structure by making the stiffness variable.

The hybrid system combines the merits of the active and passive control systems and utilizes the best features of both. Figure 4 is a conceptual drawing of the seismic response controlled structure which uses auxiliary masses of three control types, namely, the passive, active and hybrid. The differences among these types are also shown.

4. Active Seismic Response Control System

In Japan, full scale research into practical application of the active seismic response control system commenced in 1985. The history of active control system does not go back too far, and yet, in the past few years the research in this field has been rapidly expanding. This was induced only because the author stressed its importance.

The R & D process of the seismic response control is divided into its initial philosophy and five subsequent steps, as shown in Figure 5. This indicates that current development in industrial companies and various research institutes is completely uncoordinated. However, as in any new technological development, the initial concept must be clearly understood before development is commenced. In particular, as this concept was not an extension of conventional earthquake resistant design philosophy, the trail blazer had to be highly motivated as a pioneer in the field.

Needless to say, careful study and experiments are necessary to successfully combine theory and practical application. Above all, the hardware requirements in the practical application are all specific to the respective seismic response control device, and they must all be verified by experiment and resolved before success can be attained.

Therefore, this paper reports on the research up to Step 3 of Figure 5, which is to conduct a control experiment in the laboratory. The algorithm which is discussed later is also referred to in the analytical research of Step 1.

As shown in Figure 3, active seismic response control systems can be roughly categorized into the seismic response control force type and the nonresonant type. One of the former is the auxiliary mass type, which can be represented by the active mass driver system. Several institutes are presently conducting research on the auxiliary mass type. The other is the supplementary brace type as represented by the active tendon system. Research on the active tendon system is currently being conducted jointly with New York State University. In the non-building field, there is a trial for application to a Metropolitan Expressway. They are reviewed in the paragraphs below.

4.1 Control force type seismic response controlled structure

The seismic response controlled structure of the control force type provides a control force to the structure by operating the auxiliary mass installed in the structure by means of an actuator, thus reducing the response of the structure to earthquakes and strong winds. There are various systems, e.g. Active Mass Damper (AMD, Figure 6)[9], Active Mass Driver (AMD, Figure 7)[8], Active Dynamic Vibration Absorber (DVA, Figure 8)[10] and Active Damping Control (ADC, Figure 9). Some of these have already been applied to buildings. Others are in the experimental stage with the intention of future application. In the non-building structures, Active Tendon system (Figure 10)[19] was applied on an actual viaduct on the Metropolitan Expressway. The problem with these auxiliary mass methods is that they must be constantly prepared - even for large earthquakes that seldom occur - with an applicable energy source that can provide appropriate large control forces.

Subject Buildings
Subject structures include a 6-story experimental building, an 11-story office building, a 15-story office building (under planning) and the main tower of the Tokyo Bay Coastline Bridge. These structures are all of steel framing. Their weights range from several hundred to several thousand tons.

The subject external disturbances, in the cases of buildings, are all earthquakes of moderate intensity and strong winds, and are aimed at improving living comfort during frequently occurring external disturbances. The target is to reduce the response of the uncontrolled state by 2/3. The direction of control is determined in accordance with the shape of the particular building. Control may be in the transverse direction and the torsional direction, or in the two horizontal (transverse and longitudinal) directions. In the case of the steel bridge, the main objective is to suppress wind response.

Seismic Response Control Devices
In the earlier stage, electro-hydraulic type actuators have been mainly used as the seismic response control devices. However, with rapid progress in motor technology, large capacity AC servomotors are now under development. The merits of the servomotor are that its maintenance is simple, hydraulic tank space is not needed, and its response capability is excellent. If its power can be improved, it could become more widely used. The seismic response control devices are all placed on the top floor. The sensors are placed according to various planning requirements. In some cases they are installed only on the top floor, while in others they are distributed on intermediate floors. The weight of the auxiliary mass ranges from 0.5 to 1.0% of the building's weight, and their vertical support varies. e.g. the pendulum type, and the sliding-on-rail type. In all cases, to facilitate the drive function, the vibration period of the auxiliary mass system is set at a longer period than the fundamental period of the building .

Problems for Study
The current objective of the seismic response control device of the control force type is to improve the daily comfort of the occupants of medium scale buildings in strong winds and frequently occurring earthquakes of moderate intensity. The efficiency of these structures has been verified. However, it would be difficult to apply these systems in their current form to large scale buildings for controlling severe earthquakes. If it were possible to make available an auxiliary mass weight of 0.5% to 1.0% of the entire building weight, an actuator with a large stroke and an enormous control force would be required. Furthermore, a tremendous energy source would be required for its operation. Thus, it is evident that a limit exists in the building scale and its subject external force level. At this point, a semi-active concept should be given serious consideration.

4.2 Nonresonant type seismic response controlled structure

In the nonresonant type seismic response controlled structure, the system actively controls the vibration characteristics of the building so that resonance with the continuously arriving earthquake motions can be avoided and the building's response can be suppressed. To achieve this objective, the active variable stiffness system (AVS, Figure 11)[12,13] actively controls the stiffness of the building to ensure nonresonance. The model tests of this system have already been finished and it is now in the stage of practical application.

Outline

The subject structure is composed of a three-story steel frame. It is designed to be safe during earthquakes of severe to moderate intensity and typhoon winds. The seismic response control device is a two-ended rod-type hydraulic closing cylinder (cylinder locking device). A switch valve is installed in the connecting tube that joins two separate cylinder chambers. The joint between the brace and the frame is fixed or freed by opening or closing this valve. Thus, the damping factor as well as the stiffness of the entire building can be altered. The necessary energy for this operation is only the 12V of electricity needed to operate this switch.

Special features and problems

So far, the control efficiency of the nonresonant type has been confirmed by shaking table tests, and is being trially applied to an actual building. An excellent feature of this system is that the device can operate adequately with only enough electricity for emergency sources, thus contributing to power energy saving. Therefore, unlike the seismic control force type devices, this system can operate during large earthquakes. This is confirmed to be effective against earthquake waves that contain comparatively clear specific peaks in the predominant period. Research and development is being promoted so that adequate effective nonstationary state nonresonancy can be attained also for earthquake waves that are irregular and do not have a special predominant peak. Clearly, it is important to comprehend the system's dynamic characteristics (system identification) by conducting vibration tests, as well as the verification of the control system.

5. Hybrid Control System

The passive type systems shown in Figure 3 do not require special electrical power and are therefore maintenance free and mechanically simple. However, with the base isolation structure a certain fundamental period exists and in the part of base isolation devices, large deformations can occur, which properly limit its capacity. Also, it is ineffective against strong winds. The damper type systems function effectively only after the structure begins to shake, which means it has the shortcoming of slow start-up in its effect. To supplement the deficiency of the passive systems, a trial is being made to combine this with the active control system. Provided that, depending on the ratio of active to passive portions, the existence of the hybrid system can become meaningless. Up to the present, those that have been developed or currently under research include the active tuned mass damper, which combines the tuned mass damper with the actuator, and the active base isolation system, which combines the base isolation device with the actuator. These systems are described below.

5.1 Active tuned mass damper

The method of controlling the response of structures with a passive type tuned mass damper has been applied in numerous actual cases in Japan: to observation towers, tall chimneys and main towers of bridges. However, when applied to large scale structures, the following factors must be considered. (1) An enormous auxiliary mass is needed. (2) The period of the mass damper system must be synchronized to the structure. If the period becomes unsynchronized, the vibration effect is reduced tremendously. (3) As it is based on natural vibration characteristics, it takes time to get started. On the other hand, the active vibration control device of the control force type by itself would have difficulty in controlling large scale structures.

However, with the active tuned mass damper, the merits of both features are combined and effective vibration control is improved by using a small actuator. Typical examples are the Powered Passive Mass Damper (Figure 12)[22], and the Hybrid-type Mass Damper (Figure 13)[23].

Subject structures
The subject structures of the active tuned mass dampers are large scale structures such as the main towers of long large bridges and high rise buildings, whose weights range from several tens of thousands of tons to several hundred thousand tons. In particular, the powered passive mass damper system is already planned for installation in a 70-story high rise building. The external force to be suppressed is mainly strong wind with a return period of about 5 years, and the objective is to maintain a certain level of habitant's comfort.

Vibration control device
The vibration control device consists of a tuned mass damper and a control drive unit. Tuned mass dampers include the pendulum type and the type in which a curved shaped auxiliary mass is swung over a roller. Both are set to synchronize with the natural period of the building. The weight of the auxiliary mass is about 0.5% of the structure's weight. Both control drive units use an AC servo motor, but there are two systems for transmitting the force to the auxiliary mass: the ball screw type and the rack pinion type.

5.2 Active base isolation system

The earthquake response acceleration of the building with base isolation system was reduced due to lengthening the natural period by supporting the structure on laminated rubber bearings.
In recent years, many buildings have used this system because its mechanism is quite simple. However, this isolation system often has problems regarding its limitations to relative displacement. However, demand has increased for industrial facilities and sophisticated information oriented buildings with reduced absolute acceleration response, even with allowance displacement relative to the ground. In response to this demand, active control systems are expected to be added to the base isolation structures. Examples of this development are the absolute vibration control system (Figure 14)[11] and the active base isolation system (Figure 15)[24].

Subject structures
Both systems are presently being tested in laboratories to verify their basic characteristics, but no plans are yet available for application to specific buildings. However, their anticipated future use is in ultraprecise fabrication plants and sophisticated information oriented buildings. Also, under consideration is their application in important heavy buildings such as nuclear fusion reactor plants.

5.3 Problems for Study

For high rise buildings, there may be cases in which the external wind forces exceed the assumed seismic forces in structural design, depending not only on their weight, but also on their shape. Therefore, it is important to plan on response winds load by the means of passive control only, and experiments verify that enough vibration control effect can be obtained. However, as the subject building has long natural period, the devices will not commence effective operation until the building has reached a state of considerable vibration. Therefore, they are ineffective in earthquakes whose main seismic wave components (principal shock) appear in a short period of time, making them unsuitable from the standpoint of seismic response controlled structures. To counter this shortcoming, various trials and means are being continued to supplement the active control methods. However, effective devices have not been developed.

6. Control Algorithm

Depending on the utilization of measured information, the control algorithms are classified as follows, 1) feedback control, 2) feedforward control, and 3) feedback and feedforward control. In feedback control, the active control forces are determined by measured feedback responses. In feedforward control, however, the active control forces are computed from measured external excitations. In feedback and feedforward control, the active control forces are regulated by measured results of both external loads and structural responses. Various control algorithms have been investigated, including optimal control, modal control, and pole assignment.

6.1 Feedback Control

The feedback control algorithm is usually used in actual systems because it does not require a term of the external load and it is easy to design the circuit and ensure stability.

The algorithms adopted in the control force type system described in 4.1 and the hybrid control system described in Section 5 are classified as feedback control. In the system, developers contrive the adopted feedback responses and evaluation of feedback gains. The control force u is expressed as,

$$u = G_b x \qquad \text{x: state vector}$$

In the Active Mass Damper (AMD) system of Figure 6, biaxial control forces are computed by the feedback gains, that correspond to the displacements and velocities at every floor containing an auxiliary mass. The optimal G_b values are evaluated by optimal control theory. In the Active Mass Driver (AMD) system shown in Figure 7, control forces in the transverse and torsional directions are evaluated by simplified feedback gains that have effectiveness equivalent to the optimal gains obtained from optimal control theory. Feedback responses are only the velocities at the center and edge of the top floor. In the active Dynamic Vibration Absorber (DVA) system installed in the 4-storied test model, feedback gains corresponding to the displacements and velocities at all floors are evaluated from optimal control theory.

The powered passive mass damper and the hybrid-type mass damper can control only one modal response, so the feedback responses are the displacements and velocities of the system-installed floor and the mass damper.

In addition to those algorithms, an active control strategy was proposed based on the wave traveling theory and experimental studies for verification were conducted using the six-storied specimen as shown in Figure 16. The idea here is that boundary conditions are controlled to establish an absorbing boundary at the top of the building by operating a control force in proportion to the absolute velocity of the top floor.

6.2 Feedback and Feedforward Control

Various types of feedback and feedforward control are investigated, mainly by numerical analyses. In general, the control force u of this type can be expressed as,

$$u = G_b x + G_f y \qquad \text{x: state vector} \\ \text{y: external disturbance (ground motion)}$$

Yamada and Iemura [16] described that the future ground motion can be predicted as an output of one-dgree-of-freedom under a white noise input, using Kalman filtering technique and then the structural response is predicted, when the earthquake ground motion has a predominant frequency. Feedforward gain is multiplied to the predicted ground motion and feedback gain is multiplied to the predicted response to obtain the optimal control force.

p 8

Toki and Sato [15] proposed a instantaneous closed-open optimal control algorithm derived by minimizing the sum of the quadratic time-dependent performance index and the seismic energy input to the structural system. Since only the responses and observed ground motions up to present were used in this algorithm, the effect of time delay to apply the control force was a problem and also investigated. They emphasized that the control law provides feasible control algorithms that can easily be implemented for applications to seismic-excited structures.

Inoue [17] proposed a step optimal control algorithm in which the optimal control force is evaluated from the time-dependent performance index that consists of response of the structure and external load at a certain step in the discrete time domain. That is a similar method to the instantaneous optimal control algorithm.

Kawahara [18] investigated a solution algorithm for the tracking problem of discrete-time linear quadratic control based on dynamic programming to evaluate the feedforward term. In the solution, all earthquake loads are known at the beginning. However, this is impossible in actual control. Therefore, he also proposes a subtracking control system in which only a part of the earthquake load must be known in advance.

Shimogo [21] presents a feedforward control obtained by assuming that the shaping filter possesses the same natural frequency as a primary building in the frequency range of input disturbance. The optimal control force is obtained by gains that minimize the performance index containing the absolute acceleration responses. Furthermore, to improve reliability, a control system consisting of a feedback control and a feedforward control is synthesized.

6.3 Other algorithms

Ishimaru [20] proposes a unification method for the vibration mode of the structure by controlling the input vector of the disturbances in the equation of motion. This can be realized by moving additional masses at each floor in proportion to the displacement of each floor and making the participation function zero, except in the n-th vibration mode. The basic concept of this system was confirmed in the test equipment using the lever system.

7. Future Development

Development of the technology required to realize a seismic response controlled structure has progressed steadily, and the very first one has been realized in Japan. But, due to existing restrictions in the current Japanese building codes, the building with the seismic response control system simply offers living comfort during earthquakes of moderate intensity and strong winds. This has given the opportunity firstly, to make many observations of seismic response effects during the indicated external disturbances, and secondly, to verify the system's effectiveness from actual experiences. In fact, the control system presently being planned is going to be designed in a similar philosophy. The final purpose of the system, however, can be truly described as realization of a structure that will not suffer any large damage, and will not only be able to maintain the building's function, and will also prevent panic by its occupants during severe earthquakes. To attain this final target, it will probably be necessary to pass through several higher hurdles.

Essentially, the selection of the most appropriate devices and systems depends on the type of external disturbance, the type of subject structure and the aim of control. In particular, to realize the final purpose of safety and function integrity of the building structure in large earthquakes, innovative ideas are required to combine the high capacity energy absorber with the highly efficient active seismic response control system of the energy saving type.

Before such a system can be realized, several factors must be debated. If the passive control system alone can effectively suppress the responses of building structures to unknown ground motions of future earthquakes, there would be no need to expend our effort in R & D of an active control system.

However, one thing that we must not forget is the terrible disaster of the collapse of tall buildings in the 1985 Mexico Earthquake. Here, the inconceivable collapse of buildings clearly resulted because of the repeating durational resonance of the predominant period of the earthquake and the fundamental natural period of the buildings that collapsed. An important point to remember is that no one expert was able to predict earthquake ground motions of this type and then that it could cause such terrible damage. Unfortunately there are still many other unknown factors of earthquake phenomena and they are factors that will not be easily explained in the near future.

Therefore, the cause and effect relations of a disaster caused by future earthquake ground motions and the type of damage that would be suffered by building structures and resulting unfortunate conditions cannot be accurately predicted in the near future. From this viewpoint, the probability of a passive control system becoming non-functional, that is, of the system not being able to avoid a nonstationary, nonresonant state, depending on the characteristics of the ground motion, may be small, but we must bear in mind that it is not zero. We should not disregard this fact. Even though human ability is insufficient to control natural phenomena, for example, seismic activity at the earthquake focus region epicenter, we as structural engineers must not relax in any efforts to find procedures or methods to guarantee that buildings are earthquake-safe. From this viewpoint, although research into the active control system cannot immediately reach the point where the system can perform to 100% of its potential, and there remained social and economic problems to be solved, we must maintain our efforts. This workshop should greatly assist in promoting these efforts.

Figure 17 shows the outline of scheme on the active seismic response control systems. Active control systems can be categorized into two types: A) the type of seismic response control force and B) the type of nonresonant control. Both these types are being developed using a variety of ideas. Such systems that have already been developed include the AMD, Active Tendon, Acive Variable Damping and Stiffness systems. The AMD system has been verified by the building with the first installation of a fully active system in Tokyo. However, if the scale of the structure is large and the earthquake is of large magnitude, the amount of energy required to control is very large. Thus, there are problems in applying the AMD system to this kind of the large scale structure. Therefore, we should opt for a variable damping capacity or a variable stiffness system, both of which are nonresonant control type and energy-saving system. We are also considering a TMD system, which only provides active operation at the beginning of earthquakes. This system may be considered as a semi-active system.

It is important for us that we set aside energy-saving measures and economic considerations for the time being and concentrate on pure engineering considerations to research and develop such a system that is effective in earthquakes of all over the magnitudes from small to large. Only after these steps in R&D have been taken, we should consider energy saving, economics and simplification to achieve a practical system. That is the principle of research, development and implementation.

We do not doubt that, with the development of advanced technologies such as computer science technologies, an active response controlled structure will prove to be the ultimate and perfect aseismic designed structure in the near future.

<References>
1. G.W.Housner, T.T.Soong and S.F.Masri, 'An Overview of Active and Hybrid Control Research', Meeting of U.S. and Japan Panels on Active Structural Control, March 27-28, 1990, Tokyo

2. T.T.Soong, 'State-of-the-Art Review: Active Structural Control in Civil Engineering', Engineering Structures, Vol.10, pp.74-84, 1988

3. S.F.Masri, 'Seismic Response Control of Structural Systems: Closure', Proc. of the 9th World Conference on Earthquake Engineering, Vol. VIII, pp. 497-502, 1988,

4. T.T.Soong and T.Kobori, 'Outlook of Passive and Active Protective Systems for Seismic Safety of Structures', ASCE, Structures Congress, 1990, Baltimore

5. T.Kobori, 'State-of-the-Art on Dynamic Intelligent Building System', Proc. of the 8th International Congress of Cybenetics and Systems, June, 1990, New York

6. T.Kobori, 'Technology Development and Forecast of Dynamical Intelligent Building (D.I.B.)', Proc. of the International Workshop on Intelligent Structures, July, 1990, Taipei

7. T.Kobori, H.Kanayama et al., 'A Proposal of New Anti-seismic Structure with Active Seismic Response Control System -Dynamic Intelligent Building-', Proc. of the 9th World Conference on Earthquake Engineering, August 1988, Kyoto

8. T.Kobori et al., 'Study on Active Mass Driver (AMD) System -Active Seismic Response Controlled Structure', 4th World Congress of Council on Tall Buildings and Urban Habitat, Nov. 1990, Hong Kong

9. S.Aizawa, Y.Hayamizu et al., 'Study on Active Dual Axis Mass Damper', Proc. of AIJ, Annual Meeting, 2431-2433, Oct. 1990 (in Japanese)

10. T.Shimogo, K.Yoshida, T.Suzuki et al., 'Active Vibration Control System for High-rise Buildings', Proc. of AIJ, Annual Meeting, 2434-2436, Oct. 1990 (in Japanese)

11. M.Kageyama, A.Nohata et al., 'A Study on Absolute Vibration Control System of Structures', Proc. of the Dynamics and Design Conference, July 1990, Kawasaki (in Japanese)

12. T.Kobori et al., 'Shaking Table Experiment of Multi-Story Seismic Response Controlled Structure with Active Variable Stiffness (AVS) System', The 8th Japan Earthquake Engineering Symposium, Dec. 1990 (in Japanese)

13. T.Kobori et al., 'Experimental Study on Active Variable Stiffness System -Active Seismic Response Controlled Structure-', 4th World Congress of Council on Tall Buildings and Urban Habitat, Nov. 1990, Hong Kong

14. A.Mita, K Shiba et al., 'Travelling Wave Control for Tall Buildings', Proc. of AIJ, Annual Meeting, Oct. 1990

15. T.Sato and K.Toki, 'Active Control of Seismic Response of Structures', Proc. of the International Workshop on Intelligent Structures, July, 1990, Taipei

16. Y.Yamada, H.Iemura, A.Igarashi, and Y.Iwasaki, 'Phase-delayed Active Control of Structures under Random Earthquake Motion', 4th U.S. National Conference on Earthquake Engineering, 1989

17. Y.Inoue, E.Tachibana et al., 'Generalized Optimal Control Algorithms for Civil Engineering Structures', Proc. of 12th Canadian Congress of Applied Mechanics 89, 770/771, 1989

18. M.Kawahara and K.Fukazawa, 'Optimal Control of Structures Subject to Earthquake Loading Using Dynamic Programming', Proc. of JSCE No.404/I-11, April 1989

19. K.Yahagi and K.Yoshida, 'An Active Control of Traffic Vibration on the Urban Viaducts', Proc. of JSCE, No.356/I-3, April 1985 (in Japanese)

20. S.Ishimaru, et al., 'A Mode Control Method for Structure with Active-Control Devices and/or Passive-Control Ones', Journal of Structural Engineering, Vol.36B, March 1990 (in Japanese)

21. T.Shimogo, K.Yoshida et al., 'Optimal Active Dynamic Vibration Absorber for Multi-Degree-of-Freedom Systems (Feedback and Feedforward Control Using a Kalman Fiter)', Transaction of the JSME, Vol.55, No.517, 88-1358A, September 1989 (in Japanese)

22. T.Matsumoto, H.Abiru et al., 'Study on Powered Passive Mass Damper for High-rise Building', Proc. of AIJ Annual Meeting, October 1990 (in Japanese)

23. K.Tanida, Y.Koike et al., 'Development of Hybrid-Type Mass Damper Combining Active-Type with Passive-Type', Proc. of the Dynamics and Design Conference, July 1990, Kawasaki (in Japanese)

24. N.Tsujiuchi, T.Fujita et al., 'Control of Structural Vibration by Isolation Device Using Adaptive Control', Proc. of Dynamics and Design Conference, July 1990, Kawasaki (in Japanese)

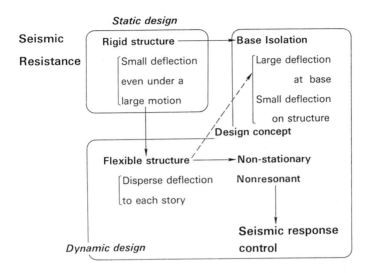

Figure 1　Relation between Conventional Aseismic Design and New Seismic Response Control Concepts

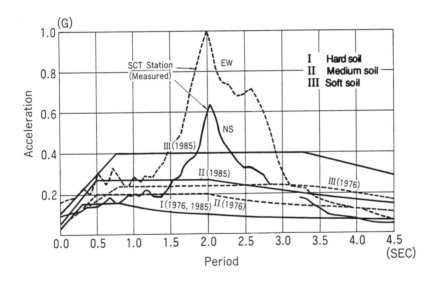

Figure 2　Response Spectra for Design and Measured Earthquake in Mexico

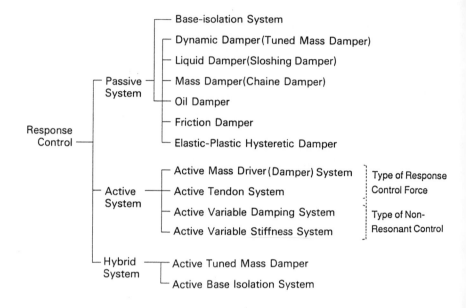

Figure 3 Application of Passive, Active and Hybrid Control System

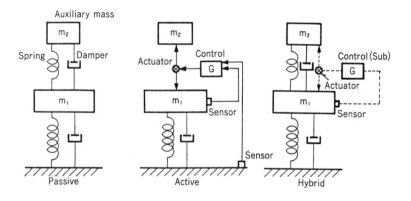

Figure 4 Model of Three-type Control System with Auxiliary Mass

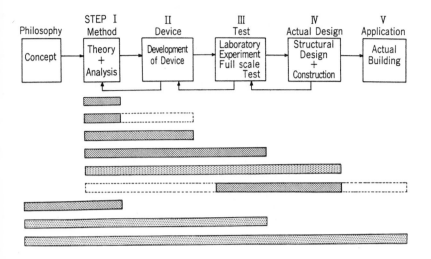

Figure 5   R & D Process on Active Control

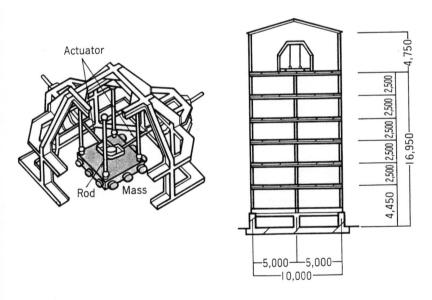

Figure 6   Active Mass Damper System [9]

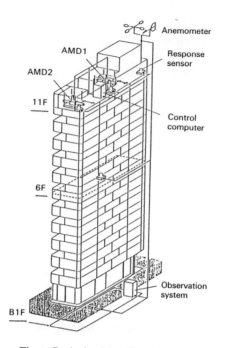

Figure 7   Active Mass Driver System [8]

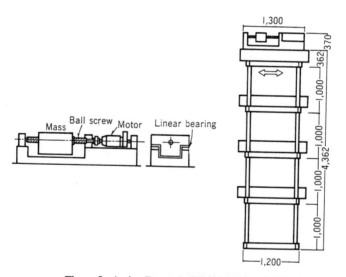

Figure 8   Active Dynamic Vibration Absorber [10]

p 16

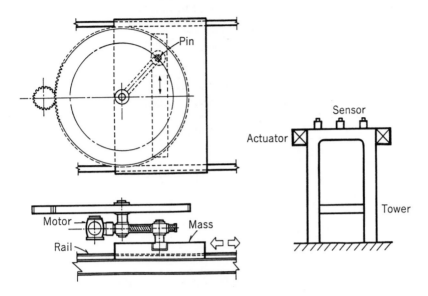

Figure 9 Active Damping Control System

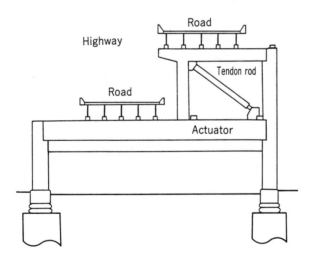

Figure 10 Active Tendon System [19]

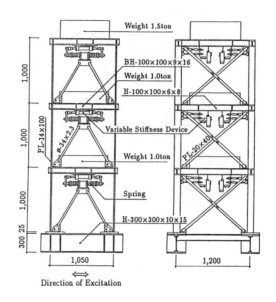

Figure 11 Active Variable Stiffness System [12,13]

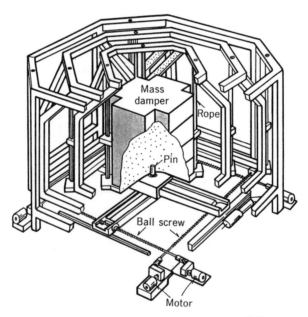

Figure 12 Powered Passive Mass Damper [22]

p 18

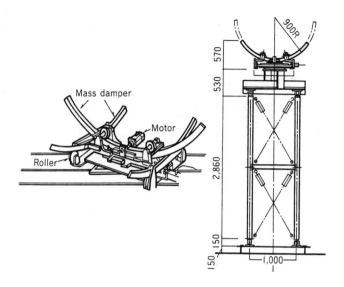

Figure 13 Hybrid-type Mass Damper [23]

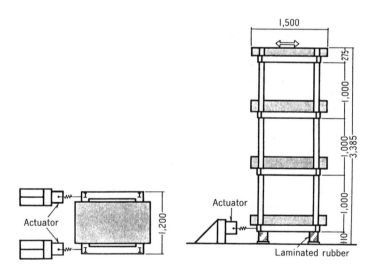

Figure 14 Absolute Vibration Contrl System [11]

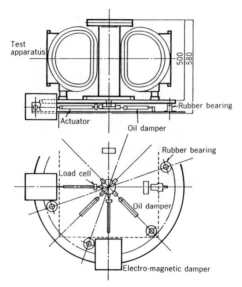

Figure 15 Active Base Isolation System [15]

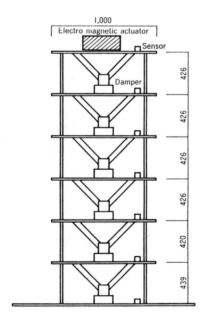

Figure 16 Frame Model for Travelling Wave control [14]

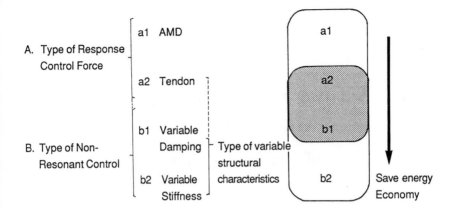

Figure 17 Outline of Scheme on Active Seismic Response Control Systems

A Survey of German and European
Activities in the Field
of Adaptive Structures

J. Melcher and E. Breitbach

German Aerospace Research Establishment (DLR)
Institute of Aeroelasticity
Structural Dynamics Division
Bunsenstr. 10, 3400 Goettingen, FRG

Phone: 011 − 49 − 551 −709 − 2342
or − 2343
Fax: − 2862

Contents

1. Introduction
2. Overview of Applications
3. Separate Activities in the Fields of
   a. Actuator / Sensor
   b. Controller
   Technologies for Adaptive Structures
4. Programs and Projects in Europe in the Field of Adaptive Structures
5. Discussion of Possible Future Directions

Areas of Activities

- aerospace systems
- traffics
- buildings
- acoustics
- optics
- measuring techniques
- manufacturing industries
- communication technologies

Applications of Active Controlled Aerospace Systems

- adaptive airplane wings
- helicopters (internal cabin, rotors)
- antennas (inflatable, deployable)
- concentrators, reflectors
- telescopes
- satellites
- solar platforms

Objectives of European Activities

- improvement of surface accuracies
- avoid of buckling
- guarantee of stability
- increase alignment precisions
- vibration suppression
- mode shape tuning
- identification of structural parameters

Subdivisions of Research
Activities in Europe

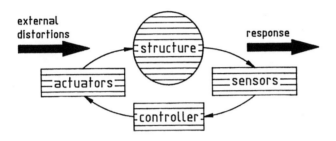

General form of the feedback loop of the actively controlled structures

In the following

1. actuator/sensor technologies

2. controller methods

3. programs and interdisciplinary projects
    covering all the four components

European Developments of Actuator and Sensor Technologies

consider

- material research
- manufacturing technologies
- measuring techniques
- advanced designs

separately in the areas of

- structural dynamics
- solid state physics
- acoustics
- mechanics
- micromechanics
- physics of flows
- thermodynamics and statistical mechanics

Piezoceramics

Material: VIBRIT®,

$$Pb\left(Zr_x Ti_{1-x}\right)O_3$$

Manufacturer: Siemens AG, Germany

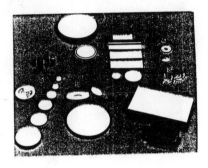

Activities

- material research
- design
- fabrication

Piezoelectric Elements

- stacks
- plates
- amplifiers
- control units

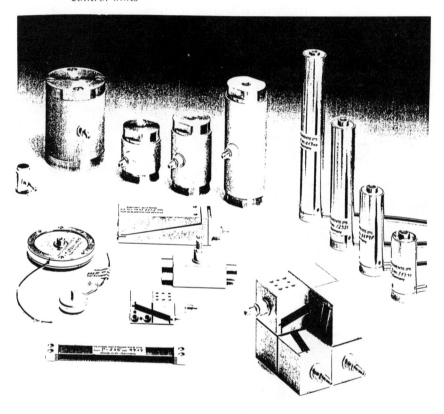

Products from Physik Instrumente, Germany

PVDF - Application   Dr. Nitsche

(A) STANDARD FOIL

Piezoelectric Constants
$d_{33} = 15\ pC/N$

$d_{31} = 8\ pC/N$

$d_{32} = 8\ pC/N$

(C) STRIPE-SENSOR

upper surface   lower surface

(B) FOIL SENSOR

Al-Coating   active   passive

(D) POINT-SENSOR

upper surface   lower surface

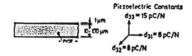

Metalli-
sierung

SENSOR   COATING   GLUE
LEADS   hard
                      soft

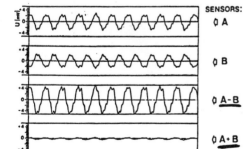

Sensor A   Sensor B

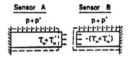

$p + p'$ $p + p'$

$\tau_o + \tau_o'$ $-(\tau_o + \tau_o')$

Separation of shear and
pressure loads.

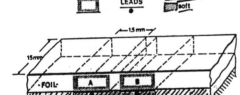

SENSORS:
◊ A

◊ B

◊ A-B

◊ A+B

$$\frac{A+B}{A-B} = \frac{(p'+\tau') + (p'-\tau')}{(p'+\tau') - (p'-\tau')} = \frac{2\cdot p}{2\cdot\tau'}$$

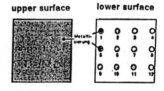

Separation test for pure shear load.

p 31

Shape Memory Alloys

Theoretical researchs:

- properties, effects
- investigation of physical models using thermodynamics and statistical mechanics

local activities:

- Prof. Müller, Technical University of Berlin

Material research and manufacturing:

- new alloys
- improvement of properties
- increase the SM-effect

local activities:

- Krupp Forschungsinstitut: cooperation with the DLR Göttingen (ARES-group)
- Prof. Hornbogen, Ruhr-University Bochum

FEREDYN

New Magnetostrictive Technology

n 1970's some alloys of rare-earth metals and iron were found to show enormous magnetostrictive effects, that is, to change their lengths when magnetized. The change can be about one hundred times greater than the effect in use in existing transducers.

Today FEREDYN has taken a lead in establishing a production process and delivering components made of these new and exciting materials.

FEREDYN magnetostrictive rods made of MAGMEK 86 ($Tb_{0.27} Dy_{0.73} Fe_{1.95}$)

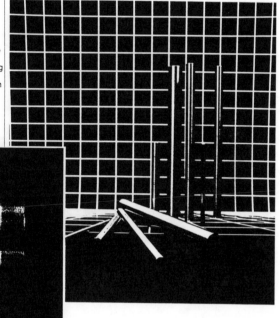

FEATURES

- Large magnetostrictive effect
- Rapid mechanical response
- Large forces available
- High acoustic power density
- High electrical—mechanical energy conversion efficiency
- Low mechanical hysteresis.
- Variable elastic modulus.

ADVANTAGES

One or more orders of magnitude greater strain than in Ni or PZT.

Frequencies of up to 10—20 kHz.

About 30 times greater power density than Ni, 10 times that of PZT.

Large coupling factor.

Low energy losses.

Application of magnetic field can increase the E-modulus by up to 60%.

Johnson Mattey GmbH

REacton™ Terfenol

Comparison with other Materials

	Rare Earth	PZT	Nickel
Maximum strain (ppm)	1500	100	40
Coupling coefficient	0.74	0.65	0.3
Density (1000 x kgm³)	9.25	7.6	8.8
Speed of sound (ms⁻¹)	2450	3100	4900

p 34

European Developments of Controller Technologies

consider

- software
- hardware
- measuring techniques

in the subdivided areas of

- computer science
- signal processing
- control theory

European Local Activities
in the Field of Control Theory

D. Guicking et al. (Germany)	Adaptive noise and vibration control algorithms, multi-channel systems, three-dimensional sound field
S. J. Elliot et al. (U.K.)	Active modal control, adaptive filtering, multichannel systems
A. E. Finzi (Italy)	Decentralized and colocated structural control units
P. Sjösten, P.Erikson (Sweden)	Adaptive noise control, real-time experiments, digital signal processors
I. D. Landau CNRS (France)	Adaptive control, algorithms, sytem identification
C. F. Ross CNRS (France)	Adaptive noise control, digital filtering, self-adaptive broadband systems
G. Grübel, J. Bals DLR (Germany)	Positioning of sensors and actuators, order reduction, optimization of a vector performance index, hyperstability theory

Activities in the Field
of Hard-/Software Developments

European Interdisciplinary Programs and Projects

of

- industries
- universities
- other research centers

DLR Research Activities

Active Flutter Suppression and Gust Load Alleviation

Targets of the research work

- Development of analytical methods for the design of
  - active flutter suppression systems
  - active gust load allevi-ation sytems
  - active stability augmen-tation

- Hardware design of the ana-lytically determined control laws and implementation of the various control systems in a quasi free-flying dynam-ically scaled aircraft model

- Simultaneous testing of all aircraft active control sys-tems in a simulated windtun-nel gust environment

Participating Nations:

Germany, France, U.K., The Netherlands

represented by

MBB/DLR, ONERA, RAE/BAe, NLR

Program: GARTEur

Period of research activ-ities

1982 - 1985

GARTEur aircraft model in the low-speed windtunnel of the DLR Göttingen

ARES
Actively Reacting Elastic Structures

Active Shape and Positioning Control
and
Active Noise and Vibration Control

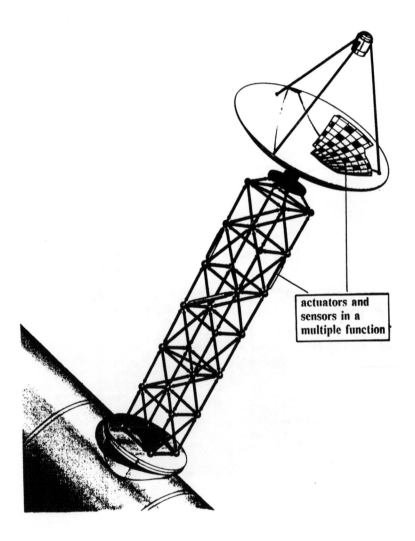

actuators and
sensors in a
multiple function

ARES
DLR - MBB Cooperation

Application: helicopters

- Active lead-lag vibration control
  (Active blade roots)

- Active vibration suppression
  (Adaptive profile)

- Higher Harmonic Control (HHC)
  (Active control rods)

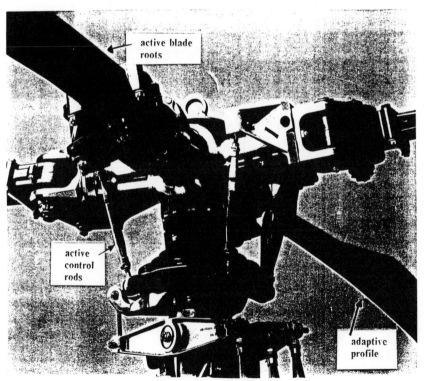

active blade roots

active control rods

adaptive profile

p 41

ARES
- further DLR Cooperations -

with

- Dornier: antennas, truss-typed structures (1990)
- VW: car carbin, motor vibrations (1989)
- NASA: contract: CSI/ARES (1989)
- Contraves (Switzerland): Space exp. with an infl. boom (1990)

Active Control
by
Metravib RDS, France
D. Duperray

Topics:

- shape and vibration control
- design methods
- utilization of new materials

Applications:

- active magnetic bearings to turbomachinery
- dynamic blocking of machinery
- active damping of truss structures

Possible Future Directions

of

European Research Activities

- materials research
  (-> utilization of new materials: magnetostrictions, liquid crystal polymers, magnetic fluids etc.)
- advanced design
  (-> new actuator/sensor types)
- digital signal processing
  (-> DSP's, neural networks, adaptive algorithms etc.)
- increase of research activities within interdisciplinary programs

References

[1] Achenbach, M.: *Simulation des Spannungs-Dehnungs-Temperatur Verhaltens von Legierungen mit Formerinnerungsvermögen.* Diss., Berlin 1987.

[2] Baier, H.: *On Optimal Passive and Active Control of Precision Spacecraft Structures.* Proc. Spacecraft Structures, CNES, Toulouse, 3-6 dec., 1985, ESA SP-238, Apr. 1986.

[3] Bals, J.: *Active Vibration Damping of Flexible Structures.* DLR-Forschungsbericht 90-03, 1090, Institute für Dynamik der Flugsysteme der DLR, Oberpfaffenhofen, Germany

[4] Duperray, B.: *Active Control.* Third International Conf. on Rotordynamics, Sept. 10-12, 1990 Lyon, France.

[5] Elliot, S.J., Curtis, A.R.D., Bullmore, A.J., Nelson, P.A. *Active minimization of harmonic enclodes sound fields.* Journal of Sound and Vibration 117 (1987)
Part 1: Theory (pp.1-13)
Part 2: A comuter simulation (pp. 15-33)
Part 3: Experimental verification (pp. 35-38).

[6] Finzi, A.E. et al.: *Active Structural Controllers Emulating Structural Elements by ICU's.* AGARD Conf. Proc. No. 397 (CP-397), Oberammergau, 9-13. Sept., 1985.

[7] Försching, H. (Editor): *Active Control Applications for Flutter Suppression and Gust Load Alleviation.* GARTEUR-TP-022 (1985).

[8] Guicking, D.: *Active Noise and Vibration Control - Reference Biliography.* 3rd Edition, 1708 citations, 1988, Drittes Physikalisches Institut, University of Göttingen.

[9] Guicking, D., Melcher, J., Wimmel, R.: *Active Impedance Control in Mechanical Structures.* Acustica,Vol. 69, 1989, pp. 39-52.

[10] Landau, J.E.: *Adaptive Control - Actual Status and Trends.* Workshop Proc. Identification and Control of Flexible Space Structures, San Diego, NASA CR 177053 Vol. 3, 1985.

[11] Melcher, J., Breitbach, E.: *New Approaches for Actively Controlling Large Flexible Space Structures.* Proceedings European Forum on Aeroelasticity and Structural Dynamics, DGLR-Bericht 89-01, paper 89-064, pp. 577-583.

[12] Melcher, J., Wimmel, R.: *Modern Adaptive Real-Time Controllers for Actively Reacting Flexible Structures.* Paper will be presented on the "1st Joint U.S./Japan Conference on Adaptive Structures, Nov. 13-15, 1990, Maui, Hawaii. Conference proceedings will be published by Technomic Publishing Company, Inc., 1991.

[13] Nitsche, W. and Weiser, N.: *Identification of Aerodynamics Flow Transitions using PVDF Foils.* Ferroelectrics, Vol. 75, pp. 339-343, 1987.

[14] Ross, C.F.: *Novel applications of active control techniques.* Proc. of the Institute of Acoustics (IOA) 8 (1986), Part 1, pp. 127-133.

Adress Correspondences

Chalmers University of Technology
Department of building acoustics
S-41296 Gothenburg, Sweden
Contact person: Mr. Sjösten

FEREDYN AB
Seminariegatan 30
S-75228 Uppsala, Sweden
Phone: 018-154596-461815-4597
Fax.: 018-154596-461815-3477
Contact person: Dr. Sullivan

German Aerospace Research Establishment (DLR)
Institute of Aeroelasticity
Bunsensr. 10
3400 Goettingen, Germany
Phone: 011-49-551-709-2343
Fax.: 011-49-551-709-2862
Contact persons: Dr. Breitbach, Mr. Melcher, Dr. Lammering, Mr. Wimmel

Johnson Mathey - Rare Earth Products
Otto-Volger-Str. 19
6231 Sulzbach/Ts. 1, Germany
Phone: 011-49-6196-70380
Fax.: 011-49-6196-73944

Krupp Forschungsinstitut
- GST -
Postfach 102252
4300 Essen 1, Germany
Phone: 011-49-201-188-1
Fax.: 011-49-201-188-2577
Contact person: Mr. Jorde

Lund University of Technology
Department of telecommunication theory
S-22007 Lund, Sweden
Fax.: 011-49-201-188-2577

MBB
Helicopter Division DV 341
P. O. B. 801140
8 Munich 80, Germany
Phone: 011-49-89-6000-3679
Contact person: Mr. Strehlow

Physik Instrumente (PI) GmbH&Co
Siemensstr.13-15
7517 Waldbronn, Germany
Phone: 011-49-7243-604-0
Fax.: 011-49-7243-604-45

Contact person: Mr. Ludwig

Ruhr University Bochum
Bochum, Germany
Phone: 011-49-234-700-5912
Contact person: Prof. Hornbogen

Siemens AG
Unternehmensbereich Kommunikationstechnik
Sicherungstechnik und Komponenten
8627 Redwitz, Germany
Phone: 011-49-9574-81-453
Contact person: Mr. Zipfel

Skalar Computer GmbH
Groner-Tor-Str. 31-32
3400 Goettingen, Germany
Phone: 011-49-551-55088
Fax.: 011-49-551-45459
Contact person: Dr. Langhans

Technical University Berlin
Institut für Thermo- und Fluiddynamik
Müller-Breslau-Str. 8
1000 Berlin 12, Germany
Contact person: Prof. Müller

Technical University Berlin
Department of Aeronautics and Astronautics
Marchstr. 14
1000 Berlin 10, Germany
Phone: 011-49-30-3142954
Contact person: Dr. Nitsche

University of Göttingen
Drittes Physikalisches Institut
Bürgerstr 42-44
3400 Götttinegn, Germany
Phone: 011-49-551-39-7727
Contact person: Dr. Guicking

State-of-the-Art of Structural Control in U.S.A.

T.T. Soong
Department of Civil Engineering
State University of New York at Buffalo
Buffalo, New York 14260

ABSTRACT

The concept of active structural control as a means of structural protection against environmental loads, developed over the last 20 years, has received considerable attention in recent years. It has now reached the stage where active systems have been installed in full-scale structures. It is the purpose of this paper to provide an overview of this development in the U.S. with special emphasis placed on laboratory experiments using model structures and on full-scale implementation of some active control systems. Included in this paper are some discussions on possible future research directions in this exciting research area.

INTRODUCTION

In structural engineering, one of the constant challenges is to find new and better means of designing new structures or strengthening existing ones so that they, together with their occupants and contents, can be better protected from the damaging effects of destructive environmental forces. As a result, new and innovative concepts of structural protection have been advanced and are at various stages of development. In the passive area, they include base isolation and a variety of other mechanical energy dissipators such as bracing systems, friction dampers, viscoelastic dampers and other mechanical devices.

Research and development of active structural control technology has a more recent origin. In structural engineering, active structural control is an area of research in which the motion of a structure is controlled or modified by means of the action of a control system through some external energy supply. In comparison with passive systems, a number of advantages associated with active systems can be cited; among them are (a) *enhanced effectiveness in motion control*. The degree of effectiveness is, by and large, only limited by the capacity of the control system; (b) *relative insensitivity to site conditions and ground motion*; (c) *applicability to multi-hazard mitigation situations*. An active system can be used, for example, for motion control against both strong wind and earthquakes; and (d) *selectivity of control objectives*. One may emphasize, for example, human comfort over other aspects of structural motion.

Thus motivated, considerable attention has been paid to active structural control research in recent years. It is now at the stage where actual systems have been designed, fabricated and installed in full-scale structures. A number of review articles [Miller et al., 1987; Kobori, 1988; Masri, 1988; Soong, 1988; Yang and Soong, 1988; Reinhorn and Manolis, 1989] and a book [Soong, 1990] have provided the reader with information and assessment on recent advances in this emerging area. In this paper, an update of this development in the U.S. is presented with special emphasis on experimental work that has been conducted in the laboratory and on full-scale implementation. This is followed by a discussion of possible future research directions in this exciting emerging technological area.

BASIC PRINCIPLES

An active structural control system has the basic configuration as shown schematically in Fig. 1. It consists of (a) sensors located about the structure to measure either external excitations, or structural response variables, or both; (b) devices to process the measured information and to compute necessary control forces needed based on a given control algorithm; and (c) actuators, usually powered by external energy sources, to produce the required forces. When only the structural response variables are measured, the control configuration is referred to as *closed-loop control* since the structural response is continually monitored and this information is used to make continual corrections to the applied control forces. An *open-loop control* results when the control forces are regulated only by the measured excitations. In the case where the information on both the response quantities and excitation are utilized for control design, the term *open-closed loop control* is used.

To see the effect of applying such control forces to a structure under ideal conditions, consider a building structure modeled by an n-degree-of-freedom lumped mass-spring-dashpot system. The matrix equation of motion of the structural system can be written as

$$M\ddot{x}(t) + C\dot{x}(t) + Kx(t) = Du(t) + Ef(t) \tag{1}$$

where M, C and K are the $n \times n$ mass, damping and stiffness matrices, respectively, $x(t)$ is the n-dimensional displacement vector, the r-vector $f(t)$ represents the applied load or external excitation, and the m-vector u is the applied control force vector. The $n \times m$ matrix D and the $n \times r$ matrix E define the locations of the control force vector and the excitation, respectively.

Suppose that the open-closed loop configuration is used in which the control force $u(t)$ is designed to be a linear function of the measured displacement vector $x(t)$, the velocity vector $\dot{x}(t)$ and the excitation $f(t)$. The control force vector takes the form

$$u(t) = K_1 x(t) + C_1 \dot{x}(t) + E_1 f(t) \tag{2}$$

where K_1, C_1, and E_1 are respective control gains which can be time-dependent.

The substitution of equation (2) into equation (1) yields

$$M\ddot{x}(t) + (C - DC_1)\dot{x}(t) + (K - DK_1)x(t) = (E + DE_1)f(t) \tag{3}$$

Comparing equation (3) with equation (1) in the absence of control, it is seen that the effect of open-closed loop control is to modify the structural parameters (stiffness and damping) so that it can respond more favorably to the external excitation. The effect of the open-loop component is a modification (reduction or total elimination) of the excitation.

It is seen that the concept of active control is immediately appealing and exciting. On the one hand, it is capable of modifying properties of a structure in such a way as to react to external excitations in the most favorable manner. On the other hand, direct reduction of the level of excitation transmitted to the structure is also possible through active control.

The choice of the control gain matrices K_1, C_1 and E_1 in equation (2) depends on the control algorithm selected. A number of control strategies for structural applications have been developed, some of which are based on the classical optimal control theory and some are proposed for meeting specific structural performance requirements. The reader is referred to Soong (1990) for discussions of some commonly used structural control algorithms.

CONTROL SYSTEMS AND EXPERIMENTAL STUDIES

As in all other new technological innovations, experimental verification constitutes a crucial element in the maturing process as active structural control progresses from conceptualization to actual implementation. Experimental studies are particularly important in this area since hardware requirements for the fabrication of a feasible active control system for structural applications are in many ways unique. As an example, control of civil engineering structures requires the ability on the part of the control device to generate large control forces with high velocities and fast reaction times. Experimentation on various designs of possible control devices is thus necessary to assess the implementability of theoretical results in the laboratory and in the field.

In order to perform feasibility studies and to carry out control experiments, investigations on active control have focused on several control mechanisms as described below.

Active Bracing System (ABS)

Active control using structural braces and tendons has been one of the most studied mechanisms. Systems of this type generally consist of a set of prestressed tendons or braces connected to a structure whose tensions are controlled by electrohydraulic servomechanisms. One of the reasons for favoring such a control mechanism has to do with the fact that tendons and braces are already existing members of many structures. Thus, active bracing control can make use of existing structural members and thus minimize extensive additions or modifications of an as-built structure. This is attractive, for example, in the case of retrofitting or strengthening an existing structure.

Active tendon control has been studied analytically in connection with control of slender structures, tall buildings, bridges and offshore structures. Early experiments involving the use of tendons were performed on a series of small-scale structural models [Roorda, 1980], which included a simple cantilever beam, a king-post truss and a free-standing column while control devices varied from tendon control with manual operation to tendon control with servovalve-controlled actuators.

More recently, a comprehensive experimental study was designed and carried out in order to study the feasibility of active bracing control using a series of carefully calibrated structural models. As Fig. 2 shows, the model structures increased in weight and complexity as the experiments progressed from Stage 1 to Stage 3 so that more control features could be incorporated into the experiments. Figure 3 shows a schematic diagram of the model structure studied during the first two stages. It is a three-story steel frame modeling a shear building by the method of mass simulation. At Stage 1, the top two floors were rigidly braced to simulate a single-degree-of-freedom system. The model was mounted on a shaking table which supplied the external load. The control force was transmitted to the structure through two sets of diagonal prestressed tendons mounted on the side frames as indicated in Fig. 3.

Results obtained from this series of experiments are reported in [Chung et al., 1988; Chung et al., 1989]. Several significant features of these experiments are noteworthy. First, they were carefully designed in order that a realistic structural control situation could be investigated. Efforts made towards this goal included making the model structure dynamically similar to a real structure, working with a carefully calibrated model, using realistic base excitation, and requiring more realistic control forces. Secondly, these experiments permitted a realistic comparison between analytical and experimental results, which made it possible to perform extrapolation to real structural behavior. Furthermore, important practical considerations such as time delay, robustness of control algorithms, modeling errors and structure-control system interactions could be identified and realistically assessed.

Experimental results show significant reduction of structural motion under the action of the simple tendon system. In the single-degree-of-freedom system case, for example, a reduction of over 50% of the first-floor maximum relative displacement could be achieved. This is due to the fact that the control system was able to induce damping in the system from a damping ratio of 1.24% in the uncontrolled case to 34.0% in the controlled case [Chung et al., 1988].

As a further step in this direction, a substantially larger and heavier six-story model structure was fabricated for Stage 3 of this experimental undertaking. It is also a welded space frame utilizing artificial mass simulation, weighing, 42,000 lbs and standing 18 ft in height.

Multiple tendon control was possible in this case and the following arrangements were included in this phase of the experiments:

(a) A single actuator is placed at the base with diagonal tendons connected to a single floor.

(b) A single actuator is placed at the base with tendons connected simultaneously to two floors, thus applying proportional control to the structure.

(c) Two actuators are placed at different locations of the structure with two sets of tendons acting independently.

Several typical actuator-tendon arrangements are shown in Fig. 4. Attachment details of the tendon system are similar to those shown in Fig. 3.

Another added feature at this stage was the testing of a second control system, an active mass damper, on the same model structure, thus allowing a performance comparison of these two systems. The active mass damper will be discussed in more detail in the next section.

For the active tendon systems, experimental as well as simulation results have been obtained based upon the tendon configurations stipulated above. Using the N-S component of the El-Centro acceleration record as input, but scaled to 25% of its actual intensity, control effectiveness was demonstrated. For example, in terms of reduction of maximum relative displacements, results under all actuator-tendon arrangements tended to cluster within a narrow range. At the top floor, a reduction of 45% could be achieved. Control force and power requirements were also found to be well within practical limits when extrapolated to the full-scale situation [Reinhorn et al., 1989].

Active Mass Damper and Active Mass Driver (AMD)

The study of this control mechanism was in part motivated by the fact that passive tuned mass dampers for motion control of tall buildings are already in existence. Tuned mass dampers are in general tuned to the first fundamental frequency of the structure, thus only effective for building control when the first mode is the dominant vibrational mode. This may not be the case; however, when the structure is subjected to seismic forces when vibrational energy is spread over a wider frequency band. It is thus natural to ask what additional benefits can be derived when they function according to active control principles. Indeed, a series of feasibility studies of active and semi-active mass dampers have been made along these lines and they show, as expected, enhanced effectiveness for tall buildings under either strong earthquakes or severe wind loads.

In the U.S., an active mass damper system was tested in conjunction with an active tendon system as described above. Using the same six-story 42,000-lb structure as shown in Fig. 4, the AMD system was placed on top of the structure, which could be operated under different conditions by changing its added mass, it stiffness and the state of the regulator. A total of 12 cases were performed in the experiment.

Extensive experimental results were obtained under various simulated earthquake excitations. A summary of results obtained under the 25% intensity El Centro excitation is given below:

Percent Reduction of Maximum Top-floor Relative Displacement:	43.3-57.2
Percent Reduction of Maximum Top-floor Acceleration:	5.5-30.7
Percent Reduction of Maximum Base Shear:	31.4-44.4
Maximum Control Force Required (kips):	0.68-2.56
Maximum Mass Peak-to-Peak Stroke (in):	3.23-10.1
Maximum Control Power Required (Kw):	0.82-5.73

One of the advantages of testing two different active systems using the same model structure is that their performance characteristics can be realistically compared. Extensive simulation and experimental results obtained based on the six-story, 42,000-lb model structure show that both AMD and ABS display similar control effectiveness in terms of reduction in maximum top-floor relative displacement, in maximum top-floor absolute acceleration, and its maximum base shear. They also have similar control requirements such as maximum control force and maximum power. Other information which may shed more light on their relative merits but is not considered here includes cost, space utilization, maintenance and other practical observations.

Pulse Generator

Pulse control has also been a subject of experimental study in the laboratory. This control algorithm was tested using a six-story frame weighing approximately 159 kg and measuring six feet in height [Miller et al., 1987; Traina et al., 1988]. Figure 5 shows the model structure together with the test apparatus which includes vibration exciter, instrumentation, pneumatic power supply, and the minicomputer used for digital control. The electrodynamic exciter, sensor, and pneumatic actuators were located at the top of the structure. The actuators consisted of two solenoids which metered the flow of compressed air at 125 psi through eight nozzles, thus generating the required control pulses.

Figure 6 shows sample measurements of the control pulse train and top-floor relative displacement when the structure was subjected to a harmonic excitation at a frequency close to the structure's fundamental frequency. It is seen that, within ten periods of onset of control, the response is reduced to approximately 15% of the uncontrolled value. The particular control law used in this experiment resulted in the straight-line decay envelope (Coulomb friction) at control initiation and exponential decay (viscous damping) at the end of control duration.

Discussions on some of the recently developed cold-gas generators having potential structural control applications can be found in [Agababian Assoc., 1984a and 1984b]. In addition, pulse control experiments involving hydraulic and electromagnetic actuators have also been conducted in the laboratory [Traina et al., 1988].

Aerodynamic Appendage

The use of aerodynamic appendages as active control devices to reduce wind-induced motion of tall buildings has several advantages, its main attractive feature being that the control designer is able to exploit the energy in the wind to control the structure, which is being excited by the same wind. Thus, it eliminates the need for an external energy supply to produce the necessary control force; the only power required is that needed to operate the appendage positioning mechanism.

For this control scheme, a wind-tunnel experiment was conducted using an elastic model at a geometric scale of roughly 1:400 [Soong and Skinner, 1981]. This is schematically presented in Fig. 7. Its stiffness was provided by a steel plate fixed at the structure core as shown, and its length was adjusted so that under planned wind conditions in the wind tunnel used in the experiment, the first mode was dominant and was observed to be approximately 5 Hz.

The aerodynamic appendage consisted of a metal plate. It was controlled by means of a 24 VDC solenoid, activated by the sign of structural velocity as sensed by a linear differential transformer, followed by appropriate carrier and signal amplifications and a differentiator. The appendage area normal to the wind direction was roughly 2% of the structural frontal area when fully extended. A boundary layer wind tunnel was used to generate the necessary wind forces.

The active control experiment was performed under various wind conditions and a peak amplitude and velocity reduction of approximately 50% was observed.

Other Control Systems

Discussed in the above are some of the most studied control mechanisms for structural applications. Many others have been proposed. Furthermore, the combined use of active-passive, or hybrid, systems have been suggested for some specific structural applications [Reinhorn et al., 1987; Kelly et al., 1988]. Hybrid control can alleviate some of the limitations which exist for either the passive system or the active system operating singly, thus leading to a very effective protective system. For example, in combination with a passive system, the force requirement of an active control system can be significantly reduced, which allows the active control device to operate at a much higher efficiency and effectiveness. At the same time, a purely passive system such as a simple elastomeric bearing is limited to low-rise structures because of the possibility of uplift of the isolator due to large horizontal accelerations. The addition of an active system is capable of minimizing this uplift effect.

Experimental research on hybrid control systems has been focused primarily on the following problems:

(a) Active control of a structure with a sliding system by reducing the frictional force between the foundation and ground [Riley et al., 1991]. This can be accomplished by connecting an active controller in the direction of motion such that the structure will slide with greatly reduced friction.

p 53

(b) Active base isolation through the use of an active controller or an active mass damper in order to create a totally vibration-free environment [Pu and Kelly, 1990, and Inaudi and Kelly, 1990].

(c) Hybrid energy dissipation where the energy dissipation characteristics are regulated during the structural response to seismic action. The regulation of energy dissipation is accomplished by utilizing active control principles [Akbay and Aktan, 1990].

While some small-scale hybrid control experiments have been carried out, its feasibility still awaits verification using more realistic model structures under realistic conditions. Cost associated with hybrid systems is another important consideration.

FULL-SCALE IMPLEMENTATION AND TESTING

As alluded to earlier, full-scale implementation of active control devices in buildings has taken place. However, all full-scale systems are currently being tested in Japan. Since two of these systems are the result of U.S.-Japan collaborations, they are included here for the same of completeness.

As a result of U.S.-Japan collaboration, an active mass damper was recently fabricated and is being tested on top of a dedicated full-scale 600-ton test structure in Tokyo as depicted in Fig. 8(a). The biaxial AMD, shown in Fig. 8(b), is of the pendulum type with a fail-safe regulator. It weighs 6 tons, approximately 1/100 of the structural weight, and has a maximum stroke of ±1.0 m with a maximum control force of 10 tons. During a recent earthquake in Tokyo, the maximum relative displacement at the top floor was observed to be 0.63 cm as compared with 2.16 cm, which is the estimated maximum value had the system not been activated.

In addition, a full-scale active bracing system has been fabricated and installed in the same dedicated test structure as described above. As shown in Fig. 9, the ABS consists of four actuators attached to bracings on the first floor. Similar to the AMD, it is designed to provide motion control in either of the two directions.

Again, the advantage of having the performance of two active systems evaluated using the same structure is obvious. In addition to providing the same base parameters for performance comparisons, this arrangement allows the calibration of the ABS and AMD systems by using one of the systems as motion inducer and the other as motion controller. Even without actual seismic motion, much of the performance characteristics can be assessed using this calibration method. During the calibration period, several feasible control algorithms can be evaluated and control parameters refined.

Performance observations of these systems under actual ground motions are being carried out by deactivating one of the systems for a period of six months in order to allow performance assessment of the other system. A total three-year observation period is planned under this activation-deactivation scheme.

FUTURE RESEARCH AND CONCLUDING REMARKS

With extensive experimental work and full-scale testing underway, active structural control research for seismic applications has entered an exciting phase. Faced with increasing demands on reliability and safety, active structural control can be an eminently logical alternative in insuring structural integrity and safety to more traditional approaches.

At the same time, however, a large number of serious challenges remain and they must be addressed before active structural control can gain general acceptance by the civil engineering and construction professions at large. Some of these issues are discussed below.

Capital Cost and Maintenance

Maintenance is certainly necessary for active systems and this is an important issue particularly due to the fact that, when active control is only used to counter large seismic and other environmental forces, it is likely that the control system will be infrequently activated. The reliability of a system operating largely in a standby mode and the related problems of maintenance and performance qualification become an important issue.

Cost, however, is not likely to be an obstacle. Recent phenomenal advances in allied technology such as computers, electronics and instrumentation all reflect favorably on the cost factor. Based on recent experiences in the fabrication of full-scale systems, active systems can in fact be more economical when used in strengthening existing structures than, for example, the use of base isolation systems. This is largely due to the fact that active systems can be designed such that they are not structurally invasive. More studies, however, are needed to address the cost issue in more concrete terms.

Reliance on External Power and Reliability

Active systems rely on power sources and, when these sources in turn rely on all the support utility systems, this power dependence on the part of an active system presents serious challenges since the utility systems, unfortunately, are most vulnerable at the precise moment when they are most needed. The scope of the reliability problem is thus considerably enlarged if all possible ramifications are considered.

It should be noted that this problem has been addressed in the design of the full-scale active bracing system discussed in the preceding section. Since the control interval for earthquake-excited motions is of the order of one minute or less for each episode, the power requirement of this system is such that it can be supplied by currently available accumulators. This design strategy would eliminate its dependence on external power at the time of control execution.

On reliability, not to be minimized is the psychological side of this issue. There may exist a significant psychological barrier on the part of the occupants of a structure in accepting the idea of an actively controlled structure, perhaps leading to perceived reliability-related concerns.

Nontraditional Technology

Since the concept of an active-controlled structure is a significant departure from traditional structural concepts, obstacles exist with respect to its acceptance by the civil engineering and construction profession at large. This is particularly true when structural safety is to rely upon an active control system. More full-scale demonstration projects are thus needed for purposes of concept verification and education.

System Robustness

As demanded by reliability, cost and hardware development, applicable active control systems must be simple. Simple control concepts using minimum number of actuators and sensors may well deserve more attention in the near future. Simple control, of course, does not mean simple problems. Since civil engineering structures are complex systems, this inherent incompatibility gives rise to a number of challenging problems from the standpoint of system robustness, controllability and effectiveness.

Active vs. Passive Control

While some progress has been made in this direction, more comprehensive studies are certainly needed in order to realistically evaluate the relative merits of alternative structural protection techniques on the basis of practical criteria such as performance, structural type, site characteristics and cost-effectiveness. However, to find answers to these questions are more long-term tasks since they will depend on specific structural applications, hardware details and a variety of other issues, many of which need to be better understood and further developed.

REFERENCES

Agababian Assoc., (1984a). *Validation of Pulse Techniques for the Simulation of Earthquake Motions in Civil Structures*, AA Rept. No. R-7824-5489, El Segondo, CA.

Agababian Assoc., (1984b). *Induced Earthquake Motion in Civil Structures by Pulse Methods*, AA Rept. No. R-8428-5764, El Segondo, CA.

Akbay, Z. and Aktan, H.M., (1990). "Vibration Control of Building Structures by Active Energy Dissipation," *Proc. of U.S. National Workshop on Structural Control Research*, Los Angeles, CA.

Chung, L.L., Reinhorn, A.M. and Soong, T.T., (1988). "Experiments on Active Control of Seismic Structures," *ASCE J. Eng. Mech.*, Vol. 114, pp. 241-256.

Chung, L.L., Lin, R.C., Soong, T.T. and Reinhorn, A.M., (1989). "Experimental Study of Active Control of MDOF Seismic Structures," *ASCE J. Eng. Mech.*, Vol. 115, pp. 1609-1627.

Inaudi, J. and Kelly, J.M., (1990). "Active Isolation," *Proc. of U.S. National Workshop on Structural Control Research*, Los Angeles, CA.

Kelly, J.M., Leitmann, G. and Soldatos, A.G., (1987). "Robust Control of Base-Isolated Structures Under Earthquake Excitations," *J. Optim. Th. Appl.*, Vol. 53, pp. 159-180.

Kobori, T., (1988). "State-of-the-art Report: Active Seismic Response Control," *Proc. Ninth World Conference on Earthquake Engrg.*, Vol. VIII, pp. 435-446, Tokyo/Kyoto, Japan.

Kobori, T., Kanayama, H. and Kamagata, S., (1988). "A Proposal of New Anti-seismic Structure with Active Seismic Response Control System," *Proc. Ninth World Conference on Earthquake Engineering*, Vol. VIII, pp. 465-470, Tokyo/Kyoto, Japan.

Masri, S.F., (1988). "Seismic Response Control of Structural Systems: Closure," *Proc. Ninth World Conference on Earthquake Engrg.*, Vol. VIII, pp. 497-502, Tokyo/Kyoto, Japan.

Miller, R.K., Masri, S.F., Dehghanyar, T.J. and Caughey, T.K., (1987). "Active Vibration Control of Large Civil Engineering Structures," *ASCE J. Eng. Mech.*, Vol. 114, pp. 1542-1570.

Pu, J.P. and Kelly, J.M., (1990). "Active Tuned Mass Damper of Base Isolated Structures," *Proc. of U.S. National Workshop on Structural Control Research*, Los Angeles, CA.

Reinhorn, A.M., Soong, T.T. and Wen, C.Y., (1987). "Base Isolated Structures with Active Control," *Proc. ASME PVD Conf.*, PVP-127, pp. 413-420, San Diego, CA.

Reinhorn, A.M. and Manolis, G.D., (1989). "Recent Advances in Structural Control," *Shock and Vibration Digest*, Vol. 21, pp. 3-8.

Reinhorn, A.M., Soong, T.T., et al., (1989). *1:4 Scale Model Studies of Active Tendon Systems and Active Mass Dampers for Seismic Protection*, Tech. Rep. NCEER-89-0026, National Center for Earthquake Engineering Research, Buffalo, NY.

Riley, M., et al., (1991). "Active Control of Absolute Motion in Sliding Systems," *Proc. of Eighth VPI & SU Symposium on Dynamics and Control of Large Structures*, Blacksburg, VA.

Roorda, J., (1980). "Experiments in Feedback Control of Structures," *Structural Control*, (H.H.E. Leipholz, ed.), North Holland, Amsterdam, pp. 629-661.

Soong, T.T. and Skinner, G.T., (1981). "Experimental Study of Active Structural Control," *ASCE J. Eng. Mech., Div.*, Vol. 107, pp. 1057-1068.

Soong, T.T., (1988), "State-of-the-art-Review: Active Structural Control in Civil Engineering," *Engineering Structures*, Vol. 10, pp. 74-84.

Soong, T.T., (1990). *Active Structural Control: Theory and Practice*, Longman, London, and Wiley, New York.

Traina, M.I., Masri, S.F., et al., (1988). "An Experimental Study of Earthquake Response of Building Models Provided with Active Damping Devices," *Proc. Ninth World Conference on Earthquake Engineering*, Vol. VIII, pp. 447-452, Tokyo/Kyoto, Japan.

Yang, J.N. and Soong, T.T., (1988), "Recent Advances in Active Control of Civil Engineering Structures," *J. Prob. Eng. Mech.*, Vol. 3, pp. 179-188.

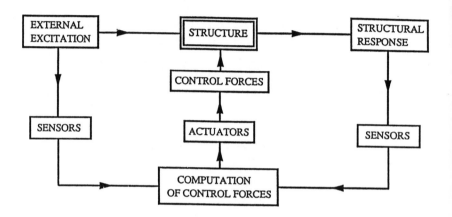

Fig. 1   Block Diagram of Active Control

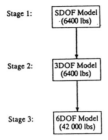

Stage 1: | SDOF Model (6400 lbs) |

Stage 2: | 3DOF Model (6400 lbs) |

Stage 3: | 6DOF Model (42 000 lbs) |

Fig. 2   Laboratory Tests of Active Bracing Systems

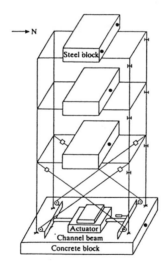

Fig. 3   Schematic Diagram of Model Structure at Stages 1 and 2

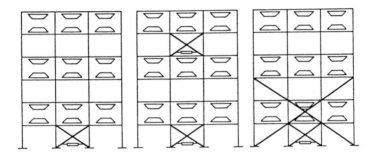

Fig. 4   Model Structure at Stage 3
and Examples of Actuator-Tendon Arrangements

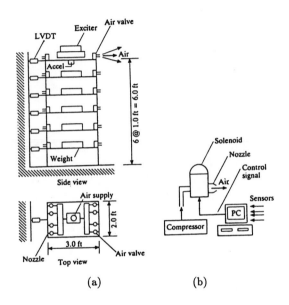

Fig. 5   Six-Story Frame with Pulse Control Mechanism
(a) Control Configuration; (b) Pneumatic Control
[Traina et al., 1988]

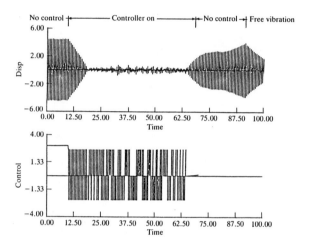

Fig. 6   Control Pulses and Top-Floor Relative Displacement

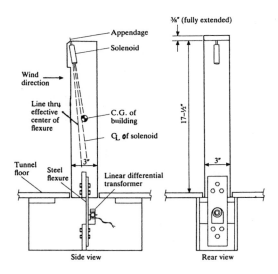

Fig. 7   Schematic Diagram of Aerodynamic Appendage

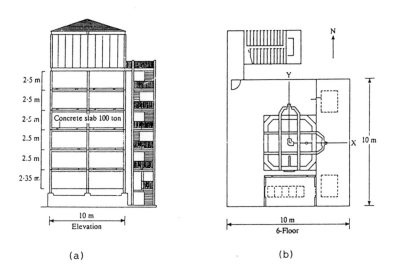

2·5 m
2·5 m
2·5 m   Concrete slab 100 ton
2.5 m
2.5 m
2·35 m.

|← 10 m →|
Elevation

(a)

N

Y

X   10 m

|← 10 m →|
6-Floor

(b)

Fig. 8   Full-Scale Dedicated Test Structure and AMD
(Courtesy of Takenaka Corporation)

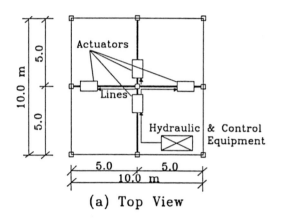

(a) Top View

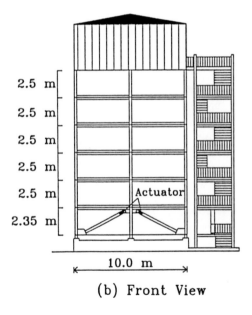

(b) Front View

Fig. 9 Active Bracing System in Test Structure

SUBMITTED PAPERS

STRUCTURAL CONTROL OF CABLE-SUPPORTED BRIDGES

by

Ahmed M. Abdel-Ghaffar, Sami F. Masri and Richard K. Miller
Department of Civil Engineering
University of Southern California
Los Angeles, California 90089-2531

SUMMARY

Cable-supported long-span bridges such as classic suspension bridges and contemporary cable-stayed brides posses little damping characteristics that help alleviate vibrations under earthquake, wind and traffic loadings. Accordingly, there is a need to implement and enhance the technology of control systems including active, passive and hybrid to absorb the energy induced in the structure under service and environmental loading conditions and to furnish more defense lines for the bridge to find its way out of the critical range of dynamic excitations. This on-going, analytical and experimental studies have the following research phases:

1. Passive Control and Seismic Isolation

In this study, guidelines and major problems encountered to passively control dynamic response are addressed through seismic isolation of these bridges using three-dimensional lead rubber bearings. An optimum seismic performance for these flexible bridges is weighed by the balance between force distribution along the bridge and tolerable bridge displacements. In order to evaluate the practicability of the isolation technique, the response of isolated bridges, at critical supports, joints and connections, is compared to non-isolated ones. Elaborate analysis for both the isolator and the bridge structure are required along with study of the impact of the different isolation parameters on the seismic performance of these long span bridges.

2. Effectiveness of Multi Active and Passive Tuned-Mass Dampers

In this preliminary study the effectiveness of passive and active structural control, for suppressing the dynamic response of cable-stayed brides subjected to earthquake excitations, is investigated. Both the dual passive tuned-mass dampers (PTMDs) and the dual active tuned-mass dampers (ATMDs) are examined and only the first symmetric and antisymmetric bending modes of bridge vibration are considered; in addition comparison with single TMD is made. Numerical example are presented to demonstrate the practical benefits and limitations of these devices in terms of damping augmentation. More research both analytically and more importantly experimental is planned for this phase of the investigation.

3. Structural Control Utilizing USC Multiple Shake Table Facility and Energy Dissipation Devices For Reduced Bridge Models

The newly acquired USC two shake-table system, at the Structural Engineering Laboratory of the Department of Civil Engineering, will be utilized to dynamically test $\frac{1}{100}$ reduced models of cable-supported bridges. The two 4×4 ft tables are separated by 12 ft (center to center) can be used to simulate differential support motions resulting from the propagation of seismic waves. In addition, because of the carefully designed system the accuracy of such two tables can be utilized to experimentally handle the dynamic control tests with reasonably accurate feedback characteristics. Thus, practical active response control concepts for these long-span and flexible structures, under earthquake excitation, can be developed. Furthermore, linear and nonlinear system identification tests for this multiple input / multiple output problem can be conducted. The generation of control action will be made through mass absorbers, actuators and "impulse" jets. Time delay effect will be investigated in the design of stability analysis of controlled distributed parameters bridges subjected concentrated control actions. And Finally, the feasibility of suing this control technique in practical application will be studied.

Experimental Study of Dual Axis Active Mass Damper

Satoru AIZAWA, Yutaka HAYAMIZU, Masahiko HIGASHINO
Yutaka SOGA and Masashi YAMAMOTO
Takenaka Technical Research Laboratory, Tokyo Japan

SUMMARY

The authors have been studying an Active Mass Damper system(AMD) which uses mass inertia as a counter force to dampen vibration of the structure affected by earthquake or wind. After small and medium test models, we constructed a full scale structure to certify the efficiency of AMD during an earthquake.

Through observation and analysis, we saw that AMD reduced the displacement response 1/2 to 1/3 that of an uncontrolled state. At the same time, acceleration response of the controlled and uncontrolled states were almost at the same level. Additionally, mass stroke and control forces calculated through analysis agree well with the observation. This proves that AMD has principally achieved its design parameters.

INTRODUCTION

As a device for use in reducing a building's sway caused by earthquakes or wind-force, we have developed an Active Mass Damper (AMD) which alters the reaction of a controlled force into an inertial force of moving mass. Following to prototype tests and medium scale tests, earthquake observation had been made using a full scale observation model to check the performance of AMD in actual earthquakes. This is a report on the behaviors of a structure equipped with AMD during small earthquakes.

OUTLINES OF AMD

An outline of this AMD is given in Figure 1. The 6-ton moving mass is suspended (T = 3.1 secs.) so that it can respond instantaneously and the responses of the structure can be controlled in two directions by electrohydraulic servo actuators which are orthogonally installed. A weight of 6 tons translates to one percent of the actual weight (600 tons) of the observation model to be described later. Based on optimal control theory, the damping control of AMD is performed by reducing the relative displacement of the structure. In order to attain this control, parameters signaled are the relative displacement and velocity between the foundation and uppermost levels, and movement of the mass. Figure 2 is a block diagram representation of the control algorithm.

RESULTS OF SHAKING TABLE TESTS

Prior to an earthquake observation using a full scale observation model, shaking table tests were implemented to verify the damping effect of AMD. As shown by Figure 3, in these experiments, laminated rubber bearings were adapted to constitute one lumped mass system. Simulating a high rise building, the design period was specified as 2 seconds, and weight as 24 tons. This analysis was implemented in the case of a controlled moving mass whose weight was 2 tons. From the free vibration test results obtained with the mass fixed, it was identified that the period was 2.08 seconds and the damping ratio was 4%. The analyses of uncontrolled conditions obtained by use of these values are shown in Fig. 4, compared to the results of the examination. In addition, controlled data are shown in Fig. 5, which indicates that the maximum displacement of the frame is reduced to about a half the above value. In controlled conditions, although there is a little difference in acceleration waveforms between the experiments and the analyses, displacement waveforms are nearly the same. The difference in acceleration waveforms should be caused by mechanical functions such as the friction force of AMD. In order to see how the time delay affects control, the analysis, obtained by specifying a time constant of 0.0187, is given in Fig. 6. In comparison with Fig. 5, no difference is observed, except that the acceleration waveforms approach those of experiments to some degree. Therefore, in these shaking table tests time delay has little influence.

OBSERVATION MODEL

Fig. 7 is the schematic drawing of a full scale model. This obervation design was modelled after the vibration character of comparatively high buildings. The AMD is installed on the top floor. This structure is provided with a floor area of 10m x 10m, 6 levels of steel frame structure, and on each level the weight is 100 tons, including RC slabs. Fig. 8 gives the fourier spectra on the 4th floor in micro tremors. Both charts show peaks clearly over the entire spectra from the 1st to 6th modes; the 1st dominant frequencies of the model are 0.9 Hz in the X (strong) direction and 0.65 Hz in the Y (weak) direction. Fig. 9 shows the comparison between the modal shapes found by micro tremor measurement and by six lumped mass analyses. In both X and Y directions, there are good similarities between the frequencies and the modal shapes found by analyses and observation; it therefore supports the validity of our analytic model proven by micro tremor level. Figs. 10 and 11 give a comparison between analyzed and observed responses of the model without control at the time of Tokyo Bay earthquake on 10th August, 1989. Let the damping ratio be 1% in the X direction and 0.3% in the Y direction (the values of the analyzed maximum relative displacement figures closest to the observation) for every modal order. In the analyzed waveforms under the conditions described above, some phase shifts are observed in the latter part. However, there is much similarity between analyses and observation in the waveforms and transfer function shapes.

EARTHQUAKE OBSERVATION

Using a full scale observation model, the control effect of AMD was examined during the Izu-Oshima earthquake on 14th October, 1989 and on 20th February, 1990. Fig. 12 illustrates the data of the former earthquake in both X and Y directions. Additionally, analyzed responses, obtained when AMD is not activated, are given in this figure. In comparison with responses in an uncontrolled state, it was observed that the relative displacement was reduced by 1/2 to 1/3 in those of a controlled state. Although the controlled acceleration responses are combined with the components of high-frequency, the maximum values are almost the same as those of the uncontrolled state. This proves that the displacement response of the structure can effectively be reduced without causing acceleration to increase with AMD. Fig. 13 shows X direction data of thelatter earthquake; it reenforces the effect obtained above. Fig. 14 gives transfer functions of both earthquakes. It proves that, when the frequency is lower than about 2 Hz, the damping effect is greater in comparison with theuncontrolled transfer functions of those observed on 10th October, 1989. However, in ranges over 3 Hz, an uncontrolled peak cannot be completely reduced. The X direction in particular, shows higher values than those in an uncontrolled state when the range is higher than 5 Hz. In this range, the higher mode is more frequent in comparison with those of the Y direction; there is a possibility that the means of set point control parameter, phase delay in control and or friction cause some excitation. These are the problems yet to be solved.

SIMULATION ANALYSES OF A FRAME STRUCTURE WITH AMD

Figs. 12 and 13 illustrate the results of simulation, as well as observation data. Both observed displacement responses and controlled currents are almost the same as the analyzed results. This proves that AMD has principally achieved its design control. The comparison of acceleration responses with the experimental model data is given by Fig. 13, based on data recorded on 20th February, 1990. Although the waveforms combine more high-frequency components in comparison with the simulated data, the functions of low-frequency components are almost identical. It may be a result of the same unidentified cause as the one described in the transfer functions. With regard to the stroke of the cylinder, it is almost identical. However, the waveforms observed do have some waviness; it will be necessary to improve the static controllability of AMD.

CONCLUSION

This report is devoted to the findings obtained through the shaking table test of a bidirectionally controlled AMD and the properties of the swaying of full scale observation model caused by a small earthquake. In a frequency range lower than 3 Hz, predicated performance can be practically obtained. However, in higher ranges, there are some problems to be solved. AMD is a device which is technically on a proper course of development, based on its stability and reliability. Therefore, our research will continue into the improvement of its performance.

REFERENCES

1) AIZAWA, S. et al, An Experiemental Study On Active Mass Damper Proceeding of Ninth WCEE, 1988, Tokyo-Kyoto, JAPAN
2) Reinhorn, A. M. et al, Experiments on Active Structural Control Under Seismic Loads, ASCE Structure Congress, 1989, San Francisco
3) Fukao, Y. et al, Experiments of Active Mass Damper Control Under Seismic Loads, ASCE Structure Congress, 1990, Baltimore

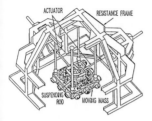

Fig.– 1 Construction of AMD

Table – 1 Specifications of AMD

Control Direction	2 (X,Y)
Weight of Mobile Mass	6tonf
Length of Suspending Rod	245cm
Stroke	95cm
Maximum Velocity	300cm／sec
Maximum Control Force	10tonf

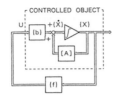

$(\dot{X}) = [A](X) + (b) U$

$U = (f)^T \cdot (X)$

$(X) = \left\{ \begin{array}{c} X \\ \dot{X} \end{array} \right\}$

(X) : STATE VECTOR

(f) : FEEDBACK COEFFICIENT
VECTOR

Fig.– 2 Block Diagram
of Control Argorithm

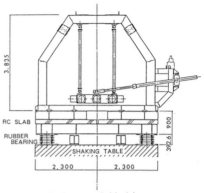

Fig.– 3 Structural Model
for Shaking Table Test

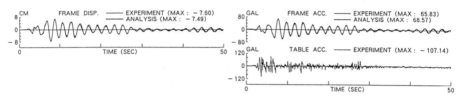

Fig.– 4 Uncontrolled Responses in Shaking Table Test

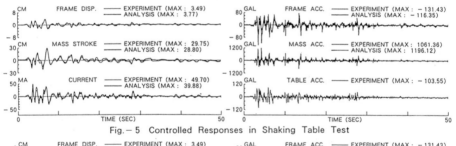

Fig.– 5 Controlled Responses in Shaking Table Test

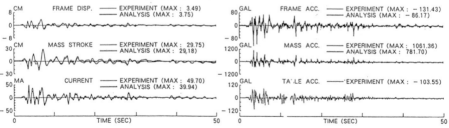

Fig.– 6 Controlled Responses Considering Time Delay in Shaking Table Test

p 71

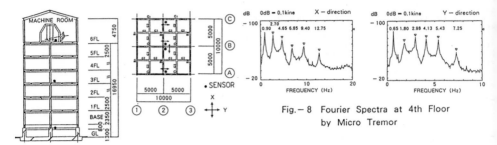

Fig.— 8 Fourier Spectra at 4th Floor
by Micro Tremor

Fig.— 7 Earthquake Observation Model

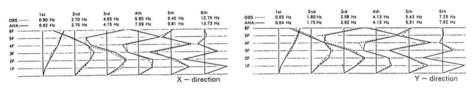

Fig.— 9 Mode Shape of Model Structure

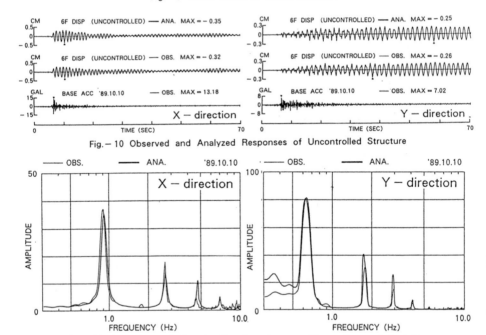

Fig.— 10 Observed and Analyzed Responses of Uncontrolled Structure

Fig.— 11 Comparison of Transfer Functions between Observation and Analysis

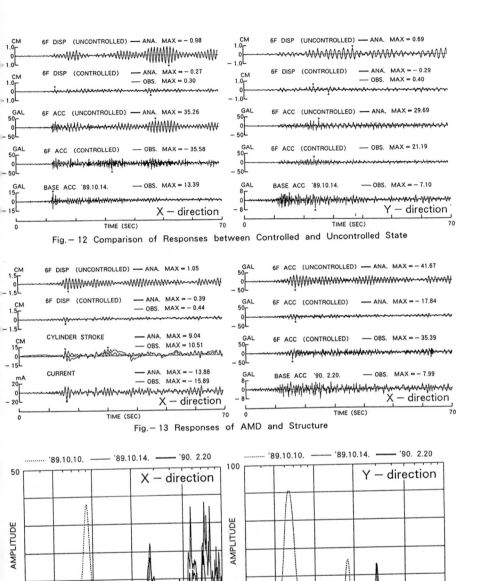

CM 6F DISP (UNCONTROLLED) —— ANA. MAX = – 0.98

CM 6F DISP (CONTROLLED) —— ANA. MAX = – 0.27
—— OBS. MAX = 0.30

GAL 6F ACC (UNCONTROLLED) —— ANA. MAX = 35.26

GAL 6F ACC (CONTROLLED) —— OBS. MAX = – 35.58

GAL BASE ACC '89.10.14. —— OBS. MAX = 13.39
X – direction

CM 6F DISP (UNCONTROLLED) —— ANA. MAX = 0.69

CM 6F DISP (CONTROLLED) —— ANA. MAX = – 0.29
—— OBS. MAX = 0.40

GAL 6F ACC (UNCONTROLLED) —— ANA. MAX = 29.69

GAL 6F ACC (CONTROLLED) —— OBS. MAX = 21.19

GAL BASE ACC '89.10.14. —— OBS. MAX = – 7.10
Y – direction

TIME (SEC)

Fig.– 12 Comparison of Responses between Controlled and Uncontrolled State

CM 6F DISP (UNCONTROLLED) —— ANA. MAX = 1.05

CM 6F DISP (CONTROLLED) —— ANA. MAX = – 0.39
—— OBS. MAX = 0.44

CM CYLINDER STROKE —— ANA. MAX = 9.04
—— OBS. MAX = 10.51

mA CURRENT —— ANA. MAX = – 13.88
—— OBS. MAX = – 15.89
X – direction

GAL 6F ACC (UNCONTROLLED) —— ANA. MAX = – 41.67

GAL 6F ACC (CONTROLLED) —— ANA. MAX = – 17.84

GAL 6F ACC (CONTROLLED) —— OBS. MAX = – 35.39

GAL BASE ACC '90. 2.20. —— OBS. MAX = – 7.99
X – direction

TIME (SEC)

Fig.– 13 Responses of AMD and Structure

········ '89.10.10. —— '89.10.14. —— '90. 2.20
X – direction

········ '89.10.10. —— '89.10.14. —— '90. 2.20
Y – direction

AMPLITUDE

FREQUENCY (Hz)

Fig.– 14 Comparison of Transfer Function in Each Earthquake

p 73

VIBRATION CONTROL OF BUILDING STRUCTURES
BY ACTIVE ENERGY DISSIPATION

Zekai Akbay[1] and Haluk M. Aktan[2]

General:

This paper describes a hybrid energy dissipation device where the energy dissipation characteristics are regulated during the structural response to seismic action. The regulation of energy dissipation characteristics are accomplished by utilizing active control principles.

The research related to energy dissipation devices are primarily based on complementing the damping capacity of the building. There are some applications include the use of energy dissipaters that are designed to be operational during the ultimate limit response state of the building [3,4]. The principle is to complement the energy dissipation capacity of the building and reduce the strength demand. The use of these energy dissipater devices are for upgrading the building structures lacking strength and/or stiffness [2]. Other conventional energy dissipation devices are to increase building damping and are not intended to supplement the structural strength [3].

Recent research by the authors led to the development of a hybrid energy dissipation device which can monitor and alter the energy dissipation characteristics of a structural system during seismic action [1]. In this implementation the energy dissipation is initiated during early stages of the seismic response and the built-up of vibratory energy is prevented. The energy dissipation of the structural system is monitored and actively corrected such that building response amplitudes are minimized. Active control principles are utilized for the monitoring and regulating the energy dissipation supply of the structure. This application was reported in recent literature by the authors under the name ASB device, and proposed for the upgrade of building structures with seismic deficiencies [1].

The active energy dissipation is a natural extension of the passive energy dissipation concept for improving seismic worthiness. The passive energy dissipation is accomplished by devices incorporated into the structural system that are designed to exhibit a stable nonlinear restoring force relationship, and dissipate energy by yielding during the interstory deformations of the building. In most instances, the device yield strength also defines the yield strength of the building story. The building design with passive energy dissipaters is based on an assumed peak demand distribution along building height during the ultimate limit state seismic motion.

---

1 Graduate Assistant, Wayne State University, Detroit, MI 48202
2 Associate Professor, Wayne State University, Detroit, MI 48202

The prototype design of the active energy dissipation device (figure 1) is based on a passive device developed by one of the authors and reported as "Friction Slip Brace". The device is a preloaded friction shaft and rigidly connected to the bracing. The operational principle is to allow the braces to deform axially through slippage along the friction interface at predefined brace loads. The slippage during seismic action dissipates significant amounts of energy. The slip load levels are designed such that brace buckling is prevented.

The passive friction devices dissipates the earthquake input energy by solid damping. The building structure behavior remains elastic until the resistive force in the bracing exceeds the friction force on the interface upon which slippage initiates. The building strength, stiffness, ductility (energy dissipation capacity) is controlled by selection of brace configuration, brace areas and slip strengths. The ultimate limit state response of the structure is elasto-plastic, non-degrading hysteretic behavior, and, the yield level is established by the slip load of the device. The building yielding is achieved by the slippage of these devices, thus the ductility is supplied without incorporating any structural damage. The device operation was verified in scaled model of a multistory building experiments on the earthquake platform [4]. Experimental results showed significant reduction of building response under ultimate limit state ground motion.

The concept described above was referred to as "Friction Braced Frames" and in this paper as "Constant Strength Friction Slip Braces", (CFSB) [4]. The problems associated with the use CFSB devices in building applications are : a) the design and operations of the devices are based on and only initiated during the limit state earthquakes, thus no functional expectation is required during serviceability limit state response, b) accurate estimation of the strength demand distribution along the building during the earthquake is essential and, c) infrequent operation of the devices are a source of concern in evaluating their reliability.

This study deals with the verifications of an actively regulated friction slip brace. The operation of the brace is controlled by the preload on the friction interface which is actively regulated based on demand during the ground motion and/or severe wind actions. These devices are kept operational at every response state to improve the serviceability under minor earthquakes and severe winds, eliminate damage under moderate earthquakes, and reduce damage during major earthquakes. The preliminary design of the device is shown in figure 1.

The active energy dissipation concept is suitable for the upgrade design of a large class of existing buildings with seismic deficiencies in addition to new buildings. The device, similar to the CFSB allows the slippage along the friction interface, however, alters building strength by actively adjusting the preload on the friction interface during building response. The objective of the actively clamped friction interface is to allow the bracing system to deform through slippage, once the preload times the friction coefficient of the interface is exceeded by the brace axial loads. The operational principle is to regulate the slip load along the friction interface during lateral displacements, thus regulating the load distribution to the bracing system for maximizing the solid damping supply under any level of seismic excitation. The slip strength selection is based on the minimization of response velocity without exceeding displacement bounds.

The implementation of the device only requires the knowledge of the enveloping structural and ground motion parameters (maximum base shear coefficent, building fundemental period and elastic response spectra estimate of the ground motion). The ASB devices will be implemented with the associated bracing as a stand-alone unit. Furthermore, the operation of the devices will be independent from each other and multiple devices will be incorporated in each story.

The advantage of the Actively Regulated Friction Slip Braces, ASB, over the Constant Strength Friction Slip Braces, CSB, and other conventional energy dissipation mechanisms, is the usability of these devices at any stage of building response. Specifically, the preload on the friction interface is adjusted actively to allow slippage in controlled amounts during all response states. The build up of the vibrational energy is dissipated by regulating the strength distribution of the building along its profile by early slippage of braces. In addition, the damage accumulation at certain stories during ultimate limit state response is eliminated by evenly dissipating the earthquake input energy while reducing the overall displacement response of the building.

Numerical Example:

To demonstrate the operation of the active energy dissipation with ASB a simple frame example is presented here. The conceptual model of the frame is shown in figure 2. Viscous damping is assumed zero in order to explicitly demonstrate the effect of the energy dissipation device on response.

Impulsive Action: The frame response is evaluated under an impulsive type ground acceleration with an amplitude of 0.15g shown in figure 3(a). The comparative displacement response of ASB and an elastic frame and the force-deformation relationship of the ASB frame is given in figure 3(b) and (c). The strength of the braces is increased incrementally starting from the zero time (figure 3(b)), thus, maximum displacement achieved by the ASB frame is greater then the elastic case. However, the early dissipation of the energy brings the frame reduces the velocity and displacement during the remaining portion of response by introducing an equivalent of 10% viscous damping to the frame. In other words, for earthquake type excitations, the input energy is dissipated immediately upon the application of the pulse to prevent buildup of vibratory energy.

Service Level Seismic Action: The frame is analyzed comparatively under a service level seismic action which is derived from the Taft earthquake record with an amplitude of 0.03g. Ground acceleration, displacement response of ASB and elastic frame and force-deformation relationship of the ASB frame is given in figure 4. The displacement response of ASB frame will be greater than the elastic response during the very early small pulses. However, as described above, the early dissipation of energy significantly reduces displacements under later large pulses.

Ultimate Level Seismic Action: The comparative response of the single story CSB and ASB under Taft 0.28g peak acceleration, is shown in figure 5. For a fair comparison the base shear strengths of both frames are taken equal to 0.25g. The ASB frame response is clearly superior and reduces building vibration amplitudes by almost 50% when compared to CSB frame.

Conclusions

The concept and development of an active energy dissipation device is presented. The preliminary building simulations indicate very favorable influence of ASB device to the reduction of response amplitudes. A preliminary design concept for the prototype device is also presented.

Further research will include multistory building response simulations, detailed design and manufacture of a prototype device and experimental verification of single device operation. Upon successful completion of the above study steps scaled building experiments on an earthquake simulator will be initiated.

References:

1. Akbay, Z., And Aktan, H. M., "Intelligent Energy Dissipation Devices," Proc. of Fourth National Conference on Earthquake Engineering, EERI, Vol. III, pp. 427-435, May 20-24, 1990, Palm Springs, California.

2. Kobori, T, "State-of-the Art Report, Active Seismic Response Control," Proc. of Ninth World Conference on Earthquake Engineering, August 2-9, 1988, Vol. VIII, Tokyo-Kyoto, Japan.

3. Pall, A. S., and Marsh, C., " Response of Friction Damped Braced Frames," Journal of ASCE, Vol. 108, ST6, June 1982, pp. 1313-1323.

4. Whittaker, A. S., Bertero, V. V., Aktan, H. M., and Giacchetti, R., " Seismic Response of a DMRSF Retrofitted with Friction-Slip Devices," Proc. EERI Annual Conference, February 9-12, 1989, San Francisco.

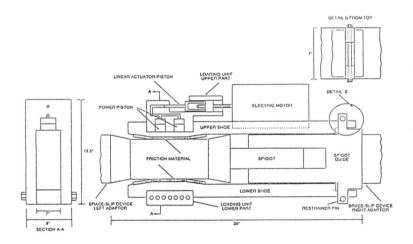

Figure 1. Preliminary design of slip device

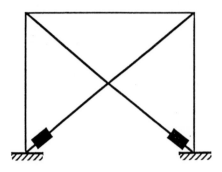

Figure 2. Example undamped frame building, T=0.5 sec.

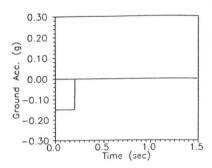

Figure 3.a. Gound acceleration

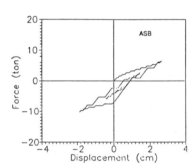

Figure 3.b. Force-deformation of ASB

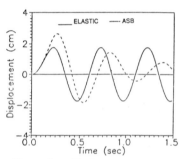

Figure 3.c. Displacement response of elastic and ASB frame.

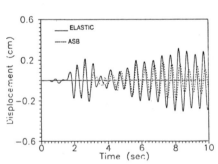

Figure 4. Displacement response of elastic and ASB frame under service level earthquake Taft3

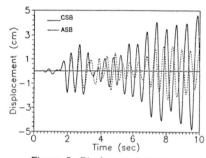

Figure 5. Displacement response of CSB and ASB frame under ultimate level earthquake Taft28

p 78

IMPLICATIONS OF MODAL TEST AND STRUCTURAL IDENTIFICATION ON ACTIVE STRUCTURAL CONTROL

A. E. Aktan, T. Toksoy, S. Hosahalli
University of Cincinnati

INTRODUCTION

General : Active structural control in civil engineering aims at controlling the responses of structures due to earthquake and strong wind forces. Different control algorithms have been investigated by many researchers. These algorithms are based on idealized system descriptions under ideal conditions and hence practical aspects need to be taken into account (Soong, 1987). The sensitivity of control to uncertainties in critical structural response mechanisms and governing parameters, realities of electro-hydro-mechanical hardware, sensors, data acquisition and computer hardware and software and the reliability of these remain important challenges. Ongoing research on active structural control has so far concentrated on idealized building physical models. The realities of actual mid rise building construction in U.S. should be considered. A very important problem in the case of physically irregular U.S. construction arises in quantifying the system parameters. A more important problem is in the need to have a complete understanding of all the critical response mechanisms of the building-foundation-soil before designing structural control. The uncertainties inherent in the existing conceptualization and analytical modeling techniques make experimental identification a necessity (Ibrahim, 87) and modal testing has emerged as a powerful tool for this.

A 27 story flat plate-core building located at the University of Cincinnati shown in Fig. 1 is used as a test specimen to explore the state-of-the-art in hardware and software for closed-loop modal testing and structural identification. The measurement of scaled modal vectors which directly transform to the lateral flexibility, including the foundation flexibility, led to developing a reliable analytical model. The serviceability-limit state force paths and deformation kinematics, including the kinematics of the foundation, are properly simulated by the model. Moreover, experience gained in conducting such a closed-loop modal test is important since similar equipment may be used in certain active structural control implementations. Understanding the behavior of hardware and software and the interaction between the components used in this test serves to identify the realities in implementations which may elude purely analytical research. The knowledge gained by this test particularly regarding the effects of exciter behavior, time lag, noise, and signal processing errors would have bearing in research and applications for active control of similar facilities.

Objective : The objective of this paper is to discuss the implications of : (a) Complete and reliable system identification based on state-of-the-art closed-loop modal testing; and, (b) performance of the hardware and software used in the closed-loop test, on active structural control.

MODAL TESTING

Previous Modal Test : Previously open-loop modal tests were conducted on the same building (Aktan, et al, 1990). Reliable data on all the critical response mechanisms could not be obtained since the signal to noise ratio afforded by the random excitation level was low. Furthermore, the accelerometers were moved from floor to floor to obtain response from all over the building and this compounded the measurement errors. The low force level (400 lbf maximum force with random excitation between 0-4 Hz) could only weakly excite the modes and the wind excitation overwhelmed the forced-excitation response. It was not possible to observe the foundation rocking and its effects on the dynamic characteristics and lateral flexibility of the building.

Description of the Current Test : The hardware and software used for the current closed-loop test are : (a) Servo Hydraulic Linear Inertia-Mass Exciter with 3500 lbs ballast; 60 HP, 30 GPM Hydraulic Pump and Pegasus 5900 Electronic Digital Servo Controller; (b) 22 PCB Model 393 C Seismic Accelerometers; (c) PCB 483B17 Amplifiers; (d) DIFA Amplifier/filter system; (e) HP 3565S Measurement Hardware system; (f) HP 9000 Series 300 Computer Work station with HP-UX operating system; (g) HP SINE Signal Processing Software (1989).

The linear inertia-mass exciter shown in Fig. 2 was preferred over rotating mass type exciter because of : (a) Faster ramp-up time; (b) quicker reaction time to frequency adjustments after ramping; (c) servo-adjustable force level during response at a given frequency; (d) capability of generating random force; and (e) smaller size and easier installation for same level of force. The 22 accelerometers were distributed throughout the building to measure the critical responses. Designing optimum instrument positioning required the knowledge gained by the previous modal test and several iterations. Swept sine excitation was used for the modal test in order to get high signal-to-noise ratios.

The closed-loop modal tests were conducted with the HP SINE signal processing software where the measurement hardware controlled swept sine excitation and acquired and processed the data in both time and frequency domains. The block diagram of the test set up is shown in Fig. 3. The advantages of closed-loop testing are : (a) Same system is used for generating the excitation signal and also for data acquisition thereby eliminating synchronization errors; (b) generating the force signal by the same system used for data acquisition enables the software to control the force level automatically; (c) decisions taken by the computer regarding settling time and signal averaging time during excitation and data acquisition stages eliminates human errors once the software is calibrated for the test by the experts.

Time Histories of Responses : The time histories of NS responses at 26th floor, obtained at the first natural frequency of the building both for the ambient wind excitation and forced excitation are compared in Fig. 4. The level of input force was sufficiently high to overcome the effects of even high-level wind excitation.

The attenuation of the response from the driving point at the 26th floor to the basement was investigated with the building being excited at its third bending mode in the EW direction. When an input harmonic force of 3200 lbf amplitude (compared to 40000 kips of estimated building weight) was given at floor 26, the driving point response was found to be 2.7 mg (thousandths of the gravitational acceleration constant), the response from the 5th floor was 1.2 mg and the basement response in the vertical direction was 0.12 mg. This indicates how much the response attenuates from the point of application of the force and this is useful in estimating the force required for a desired response at any level of the building which may have applications in the design of structural control techniques. Naturally, the attenuation characteristics would change at different limit-states of the structure. Therefore, it is desirable to conduct modal testing at the force levels corresponding to the controlled response levels. Naturally, controlled-response levels would be in the serviceability-limit-state.

Test Results : By using a higher level of force compared to the previous tests it was possible to observe the kinematics of foundation response which indicated rocking of the core foundations. The post processing of the data was done by using the "Polyreference Frequency Domain Algorithm" (Allemang and Brown, 1987). The first nine natural frequencies, damping factors and modal coefficients were obtained and the modal coefficients were further normalized to obtain unit-mass-normalized modal vectors. From the normalized modal vectors, it was possible to obtain the lateral flexibility coefficients of the building including the foundation flexibility without a need to assume inertia. The mode shapes, frequencies and damping factors obtained from both the tests for the first three bending modes in the North-South excitation direction are shown in Fig. 5. The flexibility coefficients obtained from the both the tests are compared in Fig. 6. The frequencies changed from 1.26 % to 3.60 % and the flexibility coefficients from 6 % to 13 %, due to the changes in the force type (random vs harmonic) and level in the tests.

p 80

PROBLEMS FACED, ERRORS AND IMPLICATIONS ON STRUCTURAL CONTROL

Excitation System : If an exciter is to be custom designed for a specific application, it is important to realize that when the components are put together to make up the whole system, the performance of this system may be different from the design specifications. For an electro-hydraulic linear excitation system, such as the one used in this test, the design specifications may not be realized during actual performance due to system interaction between the actuator, servo-valve and bearing hardware. The individual performance specifications of the components that make up the exciter may not provide the anticipated levels of performance when they are assembled as a system since the hydraulic-mechanical components are not rigid and have their own complex response amplifications. The hydraulic exciter used in the current modal test was custom designed to give a maximum force level of 5000 lbf (in the frequency ranges of 1.6-30 Hz), but in actual operation the exciter would safely deliver only 3200 lbf above 1.6 Hz.

An important property of the excitation systems is the ramp time. Ramp time is the time needed for the exciter to start from stationary position and reach a given force level at a given frequency. This time delay may prove especially critical in structural control implementations that rely on the use of hydraulic exciters. A delay time of anywhere between 5-10 seconds should be expected before the full performance of an exciter.

Data Acquisition System :

a) Signal Processing Errors: The response levels obtained from the tested building during forced excitation was in the order of milli g's. With such low response levels, distinguishing between noise and real response becomes a problem. In this test, the signals were averaged for 80 seconds before they were processed. During signal processing, the most time consuming but important stage is the averaging. In order to get meaningful data, signals have to be averaged for a certain amount of time to get rid of the noise content. This averaging time depends on the quality and the level of response. In the case of buildings, since usually the levels of response are low, noise caused by the elevators, people moving about, low frequency traffic noise and even the noise in the cable itself becomes a factor. For structural control applications, if the response is to be limited to mg levels, averaging time is a parameter that should be seriously taken into consideration.

b)Filtering Effect : During the test, signals from two channels, namely, the input force (reference channel) and the driving point were filtered using a low-pass filter set at a cut off frequency of 15 Hz. These channels were filtered in order to eliminate the high frequency signals generated by the exciter due to friction-slippage at the bearings during reversals. Such high-frequency signals baffled the test software. The force channel was used as the reference in calculating all the frequency response functions and the driving point channel was used in the scaling of the modal vectors to obtain the flexibility matrix. Naturally, the importance of getting accurate signals (both in terms of magnitude and phase) from these two signals was great.

Prior to filtering these channels it was known that using a low pass filter would introduce a phase lag effect. To determine this effect, the filter's characteristics was found in terms of a frequency response function of a filtered channel referenced to an unfiltered channel. This FRF indicated that the filter introduced a large amount of phase lag in the signal. Hence the need to correct for these distortions was apparent. Therefore all the FRFs obtained from the test were corrected for the filter effect by multiplying these FRFs by the filter FRF. This raw data correction was performed before any parameter estimation.

It is observed that the flexibility coefficients obtained using the force channel records corrected for the filter effect are 31.5 % less than those obtained without correcting for the filter effect. The mass normalized modal vectors obtained with correcting for filter effect are 15 % to 20 % less than those obtained without correcting for this effect. However, filtering had little effect on determining the damping factors and no effect on finding the natural frequencies. Hence, from the structural control point of view, if a channel

p 81

needs to be filtered, it is necessary to account for the filter phase lag and correct for it. If two or more channels need to be filtered then it is recommended that all channels be filtered using the same filter settings and the same kind of filter prior to any signal processing.

c) Calibration Errors : Calibrating the sensors prior to any test together with their cabling and exact amplifier settings is necessary and an important step. The drop calibration method was used to calibrate the accelerometers used in the test. The error in the calibration factors of the sensors will be directly reflected in the response level obtained from that sensor. It should be noted, however, that sensors used for monitoring the responses of a building are sensitive devices and are prone to the effects of changes in temperature, humidity and external shocks. Slight variations in calibration factors from prolonged usage should be expected and periodic re-calibration is recommended to ensure accuracy in the monitoring of the responses.

d) Connectors and Cables : Connectors and cables are the link between the sensors and the data acquisition system. Connectors often have shorts and discontinuities. A discontinuity will make that particular channel unusable whereas a short may show itself as sudden peaks in the frequency spectrum. These peaks may cause overloads and autoranging problems in the data acquisition system. The reliability of control may depend on the reliability of cables and connectors.

e) Autoranging Problem : The data acquisition system used for the modal test was a multi-channel, multiplexer system with over 2000 input channel theoretically capability. Multiplexer systems acquire data simultaneously from all channels and process them at the same time. Autoranging is a feature which adjusts the input range of the channels automatically according to the response level. Since the signal level from a channel may vary greatly with time, using autoranging will eliminate the need to adjust the input range manually and will prevent underloads and overloads from spoiling the data obtained from that channel. Naturally, not all the channels will contain the same level of response at the same time. Where some channels overload or underload others may not and since multiplexer systems have to acquire data simultaneously, whenever a channel is autoranged all the data for every channel has to be taken over again. This process can be extremely time consuming especially when long signal-averaging times are required and a large number of channels are used. Therefore, in designing a structural control technique, it is necessary to consider time-loss during autoranging and perhaps limit the number of channels needed and signal averaging time. Filtering may be an effective remedy if the correct measures of synchronized filtered data are taken.

In the modal test of the building, autoranging feature was used for only two channels out of 22. For a sweep between 0.5 Hz and 0.9 Hz with 0.015625 Hz resolution, the computer had predicted a sweep time of approximately 15 hours with all channels autoranged. To overcome this time problem, the input ranges for all but two channels were manually adjusted and fixed during the tests.

SUMMARY AND CONCLUSIONS

The need for system identification prior to the design of a structural control technique cannot be overly emphasized. Furthermore, it becomes apparent that all the hardware and software, and associated reality and reliability checks needs to be customized for a particular implementation. This is another reason for needing an accurate analytical model of the structure. Expertise-based direct analytical modeling methods for buildings have been proven unreliable if all the critical response mechanisms should be recognized and simulated. Therefore experimental identification techniques become necessary for analytical modeling in the case of consequential applications. The fact remains that even comprehensive techniques may fail to identify all the mechanisms and parameters of structure inherently as complex as actual mid-rise building-foundation-soil systems. Modal testing technique has proven helpful in identifying and quantifying the bounds of the critical mechanisms of the test building. Therefore, while many researchers have given up on dynamic testing and linearized geometric structural identification, there is definitely a need for this and the means are available.

The hardware and software used in the modal test are expected to have counterparts in structural control implementations. Hence, a complete understanding of the equipment used in modal testing and the experience gained in overcoming or realizing the possible errors and difficulties encountered during testing may shed light on the problems awaiting control implementation. Some of the parameters governing the data acquisition, signal processing and excitation signal generation are presented in this paper in a fashion that is hoped to broaden the mathematical approach to structural control techniques.

ACKNOWLEDGEMENTS

Research at the University of Cincinnati on the seismic vulnerability of buildings is partially supported by grants from the National Science Foundation through the National Center for Earthquake Engineering Research at Buffalo, New York. The contributions made during the modal testing by Drs. D. Brown and R. Allemang, SDRL of MINE Department at the University of Cincinnati are gratefully acknowledged. Invaluable contributions were made by Mr. Yoshiyuki Hashimoto of Vibration Engineering Unit, Takenaka Corporation at every stage of the test. The assistance offered by Messrs F.Deblauwe, Allyn Phillips and S. Shelley of SDRL, during the post processing stage of the data are also acknowledged. The writers would also like to thank Mr. C. Chuntavan, Ms. N. Hall and Mr. M. Zwick for their help during the test. Data acquisition and signal processing equipment was provided by SDRL, MINE Department of University of Cincinnati.

REFERENCES

Aktan, A.E., Baseheart, T.M., Shelley, S., and Ho, I.-K., "Forced-Excitation Testing and Identification of a Mid-Rise RC Building To Evaluate Vulnerability," Proceedings, Fourth U.S. National Conference on Earthquake Engineering, EERI, 1990.

Aktan, A.E., and Ho, I.-K., "Seismic Vulnerability Evaluation of Existing buildings," Earthquake Spectra, Vol. 6, No. 3, 1990.

Allemang, R.J., and Brown, D.J., "Experimental Modal Analysis And Dynamic Components Synthesis," Structural Dynamics Research Laboratory, University of Cincinnati, Cincinnati, Report No. AFWAL-TR-87-3069.

Hewlett-Packard, "HP SINE User's Manual," Software Version B.02.10, Manual No. 35631-90002, 1989.

Ho, I.-K., and Aktan, A.E., "Linearized Identification of Buildings with Cores for Seismic Vulnerability Assessment," National Center for Earthquake Engineering Research, Technical Report No. NCEER-89-0041.

Ibrahim, R.S., "Correlation of Analysis and Test in Modeling of Structures: Assessment and Review," Proceedings of workshop on "Structural Safety Evaluation Based on System Identification Approaches," edited by Natke, H.G., and Yao, J.T.P., Published by Friedr. Vieweg and Sohn, Germany, 1988.

Soong, T.T., "Active Structural Control in Civil Engineering," National Center for Earthquake Engineering Research, Technical Report No. NCEER-87-0023.

Fig. 2 Photograph of the Linear Inertia-Mass Servo Exciter.

Fig. 1 Photograph of the Test Building.

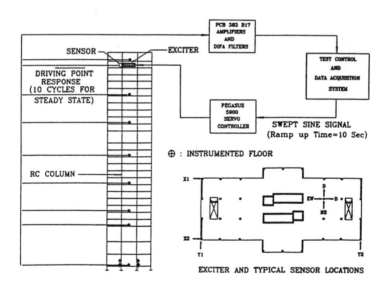

Fig. 3 Block Diagram Showing the Sensor Locations and the Test Set-Up.

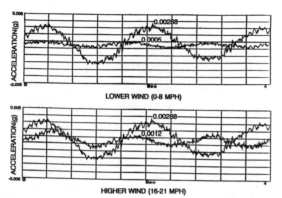

LOWER WIND (0-8 MPH)

HIGHER WIND (16-21 MPH)

Fig. 4 Time History Responses at the 26th Floor due to Different Wind Levels and Forced Excitation.

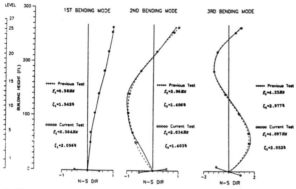

Fig. 5 Mode Shapes, Natural Frequencies and Damping for the First Three Bending Modes in the North-South Excitation Direction (Core foundation rotation amplified by 50).

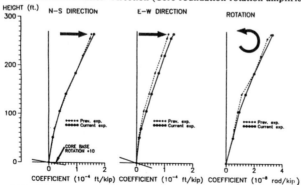

Fig. 6 Lateral Flexibility Profile of the Building When Loaded at the 26th Floor.

STRUCTURAL CONTROL AND THE E & C INDUSTRY

THOMAS L. ANDERSON

FLUOR DANIEL, INC.

INTRODUCTION

It is an honor to be invited to appear before this special group to describe examples of structural control activities at Fluor Daniel. I am dismayed that, considering the importance of the subject and the potential it holds for future E & C business in the United States, I am the only representative from industry here today. There is a message here that the workshop participants need to hear. As a minimum, I would suggest that major funding for future structural control research and applications will have to be sought from other than U.S. industry.

R & D SPENDING

The reality is that R & D spending by other countries, notably Japan, far exceeds that of the U.S. In an article in the September issue of Engineering News Record, it is reported that Japanese E & C industries spend about 0.5% of their sales on R & D, more than 10 times that of the U.S. There are upwards of 30 research institutes in Japan. This comparison paints a bleak picture for the U.S. and should send a very strong message to NSF that research in the structural control arena is simply not heavily funded by U.S. construction companies and they and cannot be counted upon to heavily invest in future R & D efforts.

APPLICATION AREAS

In the absence of a host of R & D-created market distinctives, the U.S. E & C industry has adopted an effective alternative course of action. Because the market demands that we remain sensitive to client needs, Fluor Daniel, like others in the E & C industry, is constantly working to keep abreast of new technologies; generally not marrying any one in order to create distinctive services that separate us from the rest of the industry. As a result, it appears that most of the U.S. E & C industry tends to sell their services on the basis of being a systems integrator, knowledgeable of all applicable technologies. There are some exceptions, but one suspects that our limited IRD funding has forced us into this position.

In spite of the lack of the expenditure of large sums of money on research in the U.S. in the area of construction research, there are a surprising large number of significant structural control project opportunities that the E & C industry has pursued in the U.S. To illustrate my point, the following are several examples of vibration control applications at Fluor Daniel:

o   **Base Isolation.** Construction was recently completed on the seismically isolated 911 emergency operations center for the Los Angeles County Fire Department. This is the first use of this technology for the County. We were not selected to execute the engineering of this facility on the basis of proposing to use any particular isolation hardware, or even proposing to specifically use isolation. Rather, our approach focused on our use of whatever technology made the best economic sense to meet project objectives, the most critical of which was that the facility had to remain operational during, and following, a major seismic event.

During schematic design, we demonstrated through comparative studies that seismic isolation was the preferred design solution. Isolation provided both the best means of protecting the communications equipment housed in the facility and it resulted in a six percent facility cost savings. Under other circumstances we might have achieved the same project objectives using a fixed base, hardened facility.

o   **Earth Noise.** For the last several years we have been engaged in vibration isolation studies in support of SDI. Our work has focused on the use of passive and active vibration isolation of massive underground optical bench structures. These systems are required to remain nearly motionless in an environment where background earth vibration, in the 10-20 second period range, can influence optical alignments and aiming accuracy.

o   **Ambient Vibrations.** We are routinely called upon to mitigate vibrations caused by operating machinery. Refined procedures have been developed that enable us to routinely ensure that harmful vibrations created by dynamic forces in rotating equipment and process reactor structures do not over stress foundations or influence nearby sensitive operating equipment.

o   **Semiconductor Facilities.** Computer chip manufacturing requires vibration control in order to limit ambient vibration levels in cleanrooms. For example, a recent semiconductor manufacturing facility project required a second floor cleanroom, roughly the size of a football field, to exhibit vertical vibration amplitudes of less than one micron in the 3 to 100 Hz frequency range caused by foot traffic, HVAC noise and nearby railroad traffic.

o   **Flutter**. Control of wind-induced flutter vibrations of tall, slender structures, such as stacks and flares, is routinely accomplished by the use of strakes and, on rare occasions, the introduction of passive damping elements. Strakes are rather awkward structures and in many situations extremely difficult to implement. Nevertheless, this technology has been refined to such a point that vortex shedding induced wind vibration problems are seldom encountered in our industry.

o   **Noise Control**. The reduction and control of both structural and airborne noise has become a very critical need in the power plant industry where recent projects have sited of cogeneration facilities in the near proximity of residential areas. Restrictive residential noise standards can only be met with use of heavy noise attenuation structures and rather rudimentary use of isolation technology.

The above examples deal with rather traditional vibration control applications. In response to these challenges, the E & C industry has developed rather simplistic (and generally effective) methods to solve these problems. It is suggested that several of these areas may be fertile fields for the application of active vibration control.

IMPEDIMENTS TO EXPLOITATION

There are many hurdles to overcome in the exploitation of structural control technology. Using seismic base isolation technology as an example, the following impediments to technology use can be identified:

- Lack of Codes
- Limited Research
- Absence of Education Programs
- Lack of Standardized Hardware
- Lack of Design Guides
- Benefits Not Recognized in Insurance Rates
- Liability Concerns For Peer Reviewers
- Higher Design Fees Required

Rather limited codes exist for the application, use and design of active vibration control systems. Applied research results available to the would-be practitioner are rather limited. Most practicing engineers have not taken college-level courses dealing with this topic. Active control hardware and software is all custom design; no apparent off-the-shelf equipment is available. Would-be practitioners do not have design guides to assist them in assessing the applicability of this technology. The insurance industry currently will not reduce earthquake (and other) protection premiums when these technologies are incorporated. There are numerous questions of liability for those engineers who are asked to serve as peer reviewers, a necessary step when codes and widely accepted design practices are not readily available. Finally, there is the challenge of motivation. Many in the design community would be reluctant to use new technology simply because it is new. Further, if the additional engineering required to incorporate active vibration control cannot be easily (and acceptably) passed along to the owner in the form of a higher fee, the motivation to use the technology evaporates.

WORKSHOP CHALLENGES

I challenge the workshop participants to not overlook, in their construction of a path forward, the need to spend some effort on the implementation process for the engineering community. We currently have a tough enough time as it is with passive structural control. The challenge of implementing active structural control will be even greater. So, don't forget to support the process of implementation.

I wish the workshop every success and look forward to working with you to develop your workshop products.

EFFICIENCY GLOBAL ASSESSMENT
OF PREDICTIVE CONTROL OF STRUCTURES

R.A. Andrade, J. Rodellar, F. López Almansa
Technical University of Catalonia
Barcelona, Spain

A.M. Reinhorn, T.T. Soong
State University of New York
Buffalo N.Y.

1.- Active control of structures

The active control of structures tries to minimize its response under dynamic loads by a closed loop process: at each time the response of the structure is measured and allows to a computer to calculate a control signal which is transformed into forces acting on the structure to reduce its response.

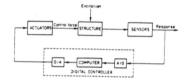

2.-Numerical simulation of the control loop

Linear equations of motion of a spatially-discretized structure:

$$M\ddot{d}+C\dot{d}+Kd=f(t)+f_c(t)$$

$f(t)$, $f_c(t)$: excitation forces and control forces

State space model:

$$\dot{x}=Fx+v+v_c$$

Where:

$$x=\begin{pmatrix}d\\\dot{d}\end{pmatrix}\text{(state vector)}\qquad F=\begin{pmatrix}0 & I\\-M^{-1}K & -M^{-1}C\end{pmatrix}$$

$$v=\begin{pmatrix}0\\M^{-1}f\end{pmatrix}\begin{matrix}\text{(excitation}\\\text{vector)}\end{matrix}\qquad v_c=\begin{pmatrix}0\\M^{-1}f_c\end{pmatrix}\begin{matrix}\text{(control}\\\text{vector)}\end{matrix}$$

$$v_c(t)=Lu(t-\tau_d)$$

$u(t-\tau_d)$ contains the control forces generated at instant $t-\tau_d$ which are applied to the structure at instant t. τ_d is the time delay in the actuators due to their physical inertia. The elements of matrix L are 0 or 1 according either to the presence or absence of actuators in every degree of freedom.

Discrete time solution (López Almansa et al. 1988):

$$x(k+1)=Ax(k)+Bu(k-d)+w(k)$$

$$A=\exp(TF)\qquad T:\text{sampling period}$$

$$w(k)=P_1v(k+1)+P_2[v(k+1)-v(k)]$$

$$P_1=F^{-1}(A-I)\qquad P_2=F^{-1}(\tfrac{1}{T}P_1-A)$$

$$B=P_1L\qquad d=\frac{\tau_d}{T}$$

3.-Predictive Control Strategy

Control force u has to be found in terms of state vector x in order to minimize the response of the structure. This strategy has been formulated in discrete-time.

1) In each sampling instant k a fictitious prediction interval [k, k+λ] is defined

$\hat{d}$: assumed delay periods

2) In the prediction scenario [k, k+λ] a model is defined to simulate the behaviour of the system. It is called **predictive model** and is used to predict, at instant k, the state in the instant $k+\lambda+\hat{d}$

$$x(k+1)=Ax(k)+Bu(k-d)+w(k)$$
(system model)

$$\hat{x}(k+j+1|k)=A\hat{x}(k+j|k)+B\hat{u}(k+j-\hat{d}|k)$$
(predictive model)

Where

$\hat{x}(k+j|k)=$ state predicted at instant k for instant k+j

$\hat{u}(\cdot|k)=$ control vectors sequency

Prediction starts from the measured value $x(k)$ of the state at instant k:

$$\hat{x}(k|k)=x(k)$$

Control forces belonging to instants previous to instant k are known:

$$\hat{u}(k-j|k)=u(k-j)\qquad (j=1,...,\hat{d})$$

The unknown control forces are assumed to be constant-shaped:

$$\hat{u}(k|k) = \dots \hat{u}(k+\lambda+1|k)$$

$$\hat{u}(k|k) = u(k)$$

3) Control force $u(k)$ is obtained by minimizing a linear-quadratic cost function J :

$$J = \frac{1}{2}\left[\hat{x}(k+\lambda+d|k) - x_r(k+\lambda+d|k)\right]^t Q\left[\hat{x}(k+\lambda+d|k) - x_r(k+\lambda+d|k)\right] + \frac{1}{2}u(k)^t R u(k)$$

Q and R are symmetric and positive-definite weighting matrices. x_r belongs to a reference trajectory which tends to zero starting from the state in instant k:

$$x_r(k|k) = x(k)$$

By imposing that the gradient of J in the direction of $u(k)$ is zeroed, one obtains (Rodellar et al. 1987):

$$u(k) = -Dx(k) - \sum_{i=1}^{\hat{d}}K_i u(k-i)$$

$D, K_1, \dots, K_{\lambda}$ are constant matrices

D = gain matrix

4) At instant $k+1$ the model is redefined from the measured state of the system $x(k+1)$.

Small values of λ provide a strong control action with an important reduction of the response.

4.-Predictive Control effect

The control action effect can be aproximated as an increase of stiffness and damping. The equation of motion of a single-degree-of-freedom controlled system is:

$$m\ddot{d} + c\dot{d} + kd = f(t) + f_c(t)$$

If no time delay is considered, predictive control may be formulated in continuous time:

$$f_c(t) = -Dx(t) = -\begin{pmatrix} D_1 & D_2 \end{pmatrix}\begin{pmatrix} d(t) \\ \dot{d}(t) \end{pmatrix} = -D_1 d(t) - D_2\dot{d}(t)$$

$$m\ddot{d} + (c+D_2)\dot{d} + (k+D_1)d = f(t)$$

$$c' = c + D_2 \qquad k' = k + D_1$$

c' and k' are, respectively, the equivalent damping and stiffness of the controlled system. Consequently, the natural frequency ω' is

$$\omega' = \sqrt{\frac{k'}{m}} \qquad T_s' = \frac{2\pi}{\omega'}$$

T_s' is the natural period of the controlled system.

5.-Predictive Control of Structures

Predictive control strategy has been considered to control structures actively. Numerical simulations (López Almansa et al. 1989), as well as control experiments on laboratory model structures (Rodellar et al. 1989) have been done. Both single and multi degree of freedom systems have been considered. All the elements in the digital control loop (discrete-time nature, time delay in the actuators, etc.) have been taken into account. Some numerical analysis have been performed.

In every single case Predictive Control has shown to be able to provide a useful and robust control action. However, an overall assessment about its efficiency has not been done. Initially it appears to be laborious since there are many parameters to deal with.

6.-Efficiency of Predictive Control of Structures

The structure is a continuous system and is modelled as a MDOF system (lumped mass, FEM, etc.). In modal coordinates (uncoupled systems) several SDOF systems are obtained. Consequently, this work is devoted to analyze a system when one degree of freedom (a mode). Two diferent analysis have been performed: plain and robustness analysis.

Topics not considered herein:

- Spillover
- Coupled systems
- Nonlinearities
- Partial measurement of the state of the system

PLAIN ASSESSMENT

All the elements in the control loop operate under ideal conditions. There are no discrepancies between the system model and the predictive model:

$$\hat{A} = A \qquad \hat{B} = B \qquad \hat{d} = d$$

ROBUSTNESS ASSESSMENT

Some of the elements in the control loop do not operate under ideal conditions. There might be missmatches between the system model and the predictive model:

$$\hat{A} \neq A \qquad \hat{B} \neq B \qquad \hat{d} \neq d$$

The following groups of parameters have to be considered:

1) Structure: M, C and K

   Modes are single degree of fredom systems

   $$m\ddot{d} + c\dot{d} + kd = f(t) + f_c(t)$$

   $$\ddot{d} + 2v\omega\dot{d} + \omega^2 d = \frac{f(t)}{m} + \frac{f_c(t)}{m}$$

   ω is the natural frequency, $T_s = 2\pi/\omega$ is the natural period and v is the damping factor. The influence of mass m is linear.

   Significant parameters : T_s, v

2) Predictive Control Algorithm: λ, $\hat{d}$, Q, R, T, x_r

$$J=\frac{1}{2}\langle\partial\quad\partial\rangle\begin{pmatrix}q_{11} & q_{12}\\ q_{21} & q_{22}\end{pmatrix}\begin{pmatrix}\partial\\ \partial\end{pmatrix}+\frac{1}{2}uRu$$

Since the main purpose of the control action is to minimize the displacement, best results are obtained when coefficients q_{12}, q_{21} and q_{22} are zero. The sampling period T is assumed to be equal to 0.01 seconds because smaller values are difficult to implement. Reference trajectory is taken zero since its effect is similar to the one of λ.

$$Q=\begin{pmatrix}1 & 0\\ 0 & 0\end{pmatrix}\quad T=0.01\ s\quad x_r=0$$

Significant parameters : λ, $\hat{d}$, R

3) Mechanical actuators. Main parameters are the time delay d, the maximum control force $|u|_{max}$ and the maximum required instantaneous power $|u\dot{u}|_{max}$.

Significant parameters : d (plain assessment)
$|u|_{max}$, $|u\dot{u}|_{max}$ (robustness assessment)

4) Excitation: Sinusoidal waves are considered in order to obtain controlled response spectra. The parameters are: the excitation period T_e, the amplitude and the total duration t_f.

Significant parameters : T_e, t_f

IMPORTANT PARAMETERS:

$$T_s, T_e, \lambda, R, \hat{d}, d$$

OTHER PARAMETERS:

$$v, t_f$$

7.-Special case. Seismic excitation

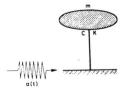

Equation of horizontal motion:

$$m\ddot{d}_r+c\dot{d}_r+kd_r=-ma(t)+f_c(t)$$

Where d_r is the relative desplacement to the ground and a(t) is the seismic acceleration.

The shear force in the basis f_b is equal to:

$$f_b(t)=f_c(t)-c\dot{d}_r-kd_r=m\ddot{d}_a$$

Where $\ddot{d}_a(t)$ is the absolute acceleration.

8.-Performance dimensionless indices

$$\Delta_1=\frac{|d_r|_{max\_control}}{|d_r|_{max}}$$

$$\Delta_2=\frac{|\dot{d}_r|_{max\_control}}{|\dot{d}_r|_{max}}$$

$$\Delta_3=\frac{|u|_{max}}{|ma|_{max}}$$

$$\Delta_4=\frac{|u\dot{u}|_{max}}{m^2|a\dot{a}|_{max}}$$

$$\Delta_5=\frac{|f_b|_{max\_control}}{|f_b|_{max}}$$

When seismic excitation, indices Δ_1, Δ_3, Δ_4 and Δ_5 are significant.

In other cases (e.g. wind loads) Δ_1, Δ_2, Δ_3 and Δ_4 are significant.

SEISMIC EXCITATION

Δ_1 : maximum relative displacement
(stresses in the members of the structure)
Δ_3 : maximum control force
Δ_4 : maximum required instantaneous power
Δ_5 : maximum absolute acceleration
(human confort and equipment safety)
maximum shear force in the base
(force transmitted to the foundation)

NON-SEISMIC EXCITATIONS

Δ_1 : maximum displacement
(stresses in the members of the structure)
Δ_2 : maximum acceleration
(human confort and equipment safety)
Δ_3 : maximum control force
Δ_4 : maximum required instantaneous power

9.-Plain assessment scheme

MAJOR PARAMETERS: T_s, T_e, λ, R, d
MINOR PARAMETERS: v, t_f

1) Controlled response spectra to obtain the optimum value of R in terms of of T_s, λ and d.

2) Controlled response spectra by considering the values of R obtained previously.

3) Assessment about the influence of other parameters: v, t_f.

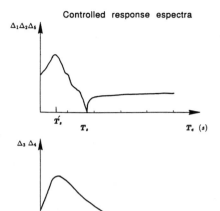

Controlled response espectra

T_e excitation period

T_s natural period

T_s' natural period of the controlled system

10.-Plain assessment conclusions

1) R. Big values of R constrain the control force u in the minimization of J and provide a smooth control action with small reduction of the response. Small values of R cause the opposite effect.

2) λ. Small values of the length of the prediction horizon imply that a fast attenuation of the response is attempted, so it results in a strong control action. The bigger is λ (greater than 5), the smoother the control action becomes.

3) d. As the time delay increases the control action looses efficiency. The effect of d is similar to the one of λ.

4) T_s. Stiff structures are difficult to control, specially if great values of λ are considered and the excitation is fast (small values of T_e).

11.-Robustnes assessment scheme

Next nonideal conditions have to be considered (López Almansa et al. 1990):

1) Structure. It might be changes in stiffness and damping (caused by damage). Adaptive techniques could be used.

2) Sensors. Measurement mistake.

- systematic error
- random error (white noise)

3) Control algorithm.

-d$\neq\hat{d}$
-poor estimation of v and T_s

4) Mechanical actuators. They may operate under saturation conditions. There are mainly two constraints:

- maximum control force $|u|_{max}$
- maximum instantaneous required power $|u\dot{u}|_{max}$

12.-Robustness assessment. Provisional conclusions

1) Control action is very robust concerning to the natural period and the damping factor estimation.

2) Control action is sensitive to discrepancies between the number of delay periods d and its assumed value $\hat{d}$.

---

p 93

13.-General conclusions

1) Predictive Control Strategy is useful to control structures actively. Control action is robust.

2) It is possible to choose the values of the Predictive Control parameters to properly control each structure. If the structure is very stiff and the excitation contains high frequencies, some instability phenomena may appear.

REFERENCES

- López Almansa F., Barbat A.H., Rodellar J., (1988) "SSP algorithm for linear and nonlinear dynamic response simulation", *International Journal for Numerical Methods in Engineering*, Vol. 26, pp. 2687-2706.

- López Almansa F., Rodellar J., (1989) "Control Systems of Building Structures by Active Cables", *Journal of Structural Engineering ASCE*, Vol. 115, pp. 2897-2913.

- López Almansa F., Rodellar J., (1990) "Feasibility and Robustness of Predictive Control of Building Structures", *Earthquake Engineering and Structural Dynamics*, Vol. 19, pp. 157-171.

- Rodellar J., Barbat A.H., Martín Sánchez, J.M., (1987), "Predictive Control of Structures", *Journal of Engineering Mechanics ASCE*, Vol. 113, No. 6, pp. 797-812.

- Rodellar J., Chung L.L., Soong T.T., Reinhorn A.M., (1989), "Experimental Digital Control of Structures", *Journal of Engineering Mechanics ASCE*, Vol. 115, No. 6, pp. 1245-1261.

ACKNOWLEDGEMENTS

This work has been partially supported by the Spanish Government (DIRECCION GENERAL DE INVESTIGACION CIENTIFICA Y TECNICA). Grant No. BE91-024.

The stay of the Spanish authors in Buffalo has been finantially supported by the State University of New York.

p 94

Point Control of Linear Distributed Parameter Structures
by
Lawrence A. Bergman[1] and D. Michael McFarland[2]
Department of Aeronautical and Astronautical Engineering
University of Illinois at Urbana-Champaign

INTRODUCTION

The rigorous analysis of passive linear discrete-distributed systems has been addressed in a recent series of publications by Bergman and Nicholson [1,2] and Bergman and McFarland [3,4]. It was demonstrated that systems of considerable complexity can be assembled by coupling to a linear distributed parameter substructure any of several types of lumped parameter mechanical subsystems and that a number of advantages arise when the distributed nature of the primary subsystem is maintained throughout the analysis in the form of a dynamic Green's function. Then the displacement of all elements of the system can be described by functions whose domain is comprised of both continuous and discrete parts, and a Hilbert space of such functions can be defined. The global distributions of stiffness and damping can be expressed in terms of hybrid differential-difference operators analogous to the differential operators that commonly arise in modeling simple distributed parameter structures, and in the hybrid space all of the usual machinery of classical modal analysis can be applied to these complex systems.

Similar though more complicated problems can be posed by extending the class of discrete subsystems to active components. Then the problem becomes one of controlling a passive distributed element via active control forces applied at discrete points. Of particular interest is the case where the control forces are derived by feeding back a linear combination of measured structural displacements and velocities. Although such systems can be made almost arbitrarily complex, a simple example will be used here to demonstrate the application of continuum methods to active structural control problems.

BACKGROUND

Consider the transverse vibration of a linear distributed parameter structure with bending stiffness and damping described, respectively, by the linear differential operators L_k and L_c in the spatial variables x_1, x_2. The nondimensional governing equation is

$$\rho w_{,tt}(\mathbf{x},t) + L_c w_{,t}(\mathbf{x},t) + L_k w(\mathbf{x},t) = f(\mathbf{x},t) + \sum_{j=1}^{S} F_j(t)\delta(\mathbf{x} - \boldsymbol{\eta}_j) \tag{1}$$

where $\mathbf{x} = \{x_1 x_2\}^T$, $w(\ )$ is the transverse displacement field, ρ is the mass distribution per unit area, $f(\ )$ is the external distrubance, $\delta(\ )$ is the Dirac delta function, and S is the number of actuators, each acting at $\boldsymbol{\eta}_j$. Further, let

$$F_j(t) = -\sum_{i=1}^{R}[g_{ij}^d w(\boldsymbol{\xi}_i,t) + g_{ij}^v w_{,t}(\boldsymbol{\xi}_i,t)] \tag{2}$$

[1] Associate Professor
[2] Graduate Research Assistant; presently Assistant Professor, Department of Mechanical Engineering, University of Connecticut

where R is the number of sensors, each located at ξ_i, and g^d_{ij}, g^v_{ij} are gains weighting the contributions of the ith displacement and velocity signals, respectively, to the jth control force.

Ideally, one would like to analyze this system with all RS feedback loops closed; however, the complications encountered are similar to those arising in the analysis of nonclassically damped passive systems due to the velocity feedback which acts as "electronic damping". Adapting the approach developed for passive systems, the free vibration response of the system less velocity feedback, passive damping and external disturbances is assumed to be separable in space and time and the spatial component is solved in terms of the Green's function $G(\ )$ of the distributed member,

$$W(\mathbf{x}) = -\sum_{i=1}^{R}\sum_{j=1}^{S} g^d_{ij} W(\xi_i) G(\mathbf{x}, \boldsymbol{\eta}_j; \alpha) \tag{3}$$

where α is an eigenparameter. Letting $\mathbf{x} \to \xi_i$, $i = 1, \ldots, R$, produces a system of scalar equations which can be written in the form

$$\mathbf{Q}(\alpha)\mathbf{w} = 0 \tag{4}$$

where

$$\mathbf{Q}_{ij}(\alpha) = \delta_{ij} + \sum_{k=1}^{S} g^d_{jk} G(\xi_i, \boldsymbol{\eta}_k; \alpha) \qquad i, j = 1, \ldots, R \tag{5}$$

and

$$\mathbf{w} = \{W(\xi_1) \cdots W(\xi_R)\}^T \tag{6}$$

The characteristic equation, roots of which are the nondegenerate system eigenvalues, follow from

$$det\ \mathbf{Q}(\alpha) = 0 \tag{7}$$

Degenerate eigenvalues, for which the determinant of $\mathbf{Q}$ need not vanish, correspond instead to $\mathbf{w} = 0$, and so degenerate system modes are modes of the distributed subsystem with nodes at all sensor locations, from which no actuator forces can be developed. Notice that actuator locations $\boldsymbol{\eta}_j$, $j = 1, \ldots, S$, enter the eigenproblem only as arguments of the Green's functions in the characteristic equation, while sensor locations ξ_i, $i = 1, \ldots, R$, comprise the arguments of the elements of $\mathbf{w}$ and also appear in the Green's functions. Hence, degenerate modes of the closed-loop system are neither observed nor controlled in such a configuration (although the modes are not strictly uncontrollable), while nondegenerate modes are observable by those sensors not at nodes and are controlled only by those actuators to which one or more of these sensor outputs are fed back.

The solution of the determinantal equation for the eigenvalues of the closed loop system α_r, $r = 1, 2, \ldots$, is a difficult task owing to the presence of multiple feedback paths, and is discussed further in the references. The α_r's are substituted into eq.(4) to obtain the eigenvectors $\mathbf{w}_r$ from which the eigenfunctions $W_r(\mathbf{x})$ are constructed. The derivation of the orthogonality relation for the eigenfunctions of the closed loop system leads to

$$(\alpha_r^4 - \alpha_s^4) \int_\Omega W_r(\mathbf{x}) W_s(\mathbf{x}) dx = \sum_{i=1}^{R}\sum_{j=1}^{S} g^d_{ij}[W_r(\xi_i)W_s(\boldsymbol{\eta}_j) - W_r(\boldsymbol{\eta}_j)W_s(\xi_i)] \tag{8}$$

It is not apparent that the RHS of eq.(8) should always vanish. Some insight can be gained by considering the simpler case of an equal number, R, of sensors and actuators, co-located ($\xi_i = \boldsymbol{\eta}_i$, $i = 1, \ldots, R$) and with $g^d_{ij} = 0$ for $i \neq j$. Such an active system behaves much like a continuous subsystem coupled to R elastic supports, and the RHS does indeed go to zero. More generally, sensors and actuators are not co-located or equal in number, and up to RS feedback paths exist among them. Thus, the modes $W_r(\mathbf{x})$ are not orthogonal in the usual sense and, in fact, the system of eq.(1) and (2) can be shown to be non self-adjoint.

Assuming the simpler case applies, the velocity feedback and passive damping are included, and the response of the closed loop system is determined by eigenfunction expansion, yielding the infinite dimensional system

$$\mathbf{M}\ddot{\mathbf{a}}(t) + \mathbf{C}\dot{\mathbf{a}}(t) + \mathbf{K}\mathbf{a}(t) = \mathbf{f}(t) \tag{9}$$

where the damping and stiffness matrices reflect the effects of displacement and velocity feedback. The elements of the matrices in eq.(9) are given by

$$\mathbf{M}_{rs} = d_s \delta_{rs}$$

$$\mathbf{C}_{rs} = C d_s \delta_{rs} + \sum_{i=1}^{R} g_i^v W_r(\boldsymbol{\xi}_i) W_s(\boldsymbol{\xi}_i)$$

$$\mathbf{K}_{rs} = \alpha_s^4 d_s \delta_{rs}$$

$$f_r(t) = \int_\Omega W_r(\mathbf{x}) f(\mathbf{x},t) dx \tag{10}$$

where where the sth modal mass d_s is defined by

$$d_s \delta_{rs} = \int_\Omega W_r(\mathbf{x}) W_s(\mathbf{x}) dx \tag{11}$$

The solution to eq.(1) is then the series

$$w(\mathbf{x},t) = \sum_{r=1}^{\infty} W_r(\mathbf{x}) a_r(t) \tag{12}$$

EXAMPLE

A common goal in structural control is to augment the damping of an existing structure through negative feedback of inertial and/or relative measured velocities. Often the primary structure is flexible and lightly damped so that active control must be used to attain acceptable response following a disturbance. Consider, here, an undamped thin square simply supported plate controlled by a single co-located actuator-sensor pair at $\boldsymbol{\xi} = (0.3\ 0.65)^T$ such that all damping is due to velocity feedback, as shown in Fig. 1. Because of the aspect ratio of the plate and its rotational symmetry, most of its open loop eigenvalues occur with multiplicity > 1, and a number of them will become degenerate with $W_r(\boldsymbol{\xi}) = 0$, as shown in Table 1. When the loop is closed, the configuration space damping matrix has elements

$$\mathbf{C}_{rs} = g^v W_r(\boldsymbol{\xi}) W_s(\boldsymbol{\xi}), \quad r,s = 1,2,\ldots. \tag{13}$$

Clearly, if the sth eigenfunction has a zero at $\boldsymbol{\xi}$, the corresponding s^{th} row and column of $\mathbf{C}$ will be null and the damping matrix will be non-negative definite which is inadequate to ensure asymptotic stability. Thus, the system is marginally stable, with components of its response due to modes which have $\boldsymbol{\xi}$ as a node being undamped, shown in Table 2. This is demonstrated in Fig. 2, which shows the response of the plate to an impulsive load applied uniformly over the surface. This serves as a simple illustration of the observability and controllability ideas discussed earlier in the paper.

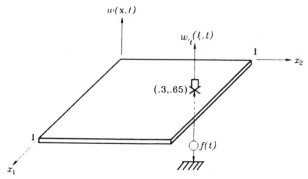

Fig. 1 - SQUARE SIMPLY SUPPORTED PLATE WITH
CO-LOCATED ACTUATER-SENSOR PAIR.

SYSTEM EIGENVALUES			
OPEN LOOP		CLOSED LOOP	
α_r	MULTIPLICITY	α_r	MULTIPLICITY
4.44288	1	4.93407	1
7.02481	2	7.02481 *	1
8.88576	1	7.43746	1
9.93458	2	9.02174	1
11.3272	2	9.93458 *	1
12.9531	2	9.95041	1
13.3286	1	11.3272 *	1
14.0496	2	11.3357	1

Table 1 - EIGENVALUES OF THE OPEN AND CLOSED LOOP
SYSTEM LESS VELOCITY FEEDBACK (* = DEGENERACY).

SYSTEM EIGENVALUES CLOSED LOOP (g^v= 10)
3.87409 ± j 26.3326
0 ± j 49.3480
29.9368 ± j 49.0399
4.86410 ± j 72.7759
.689118 ± j 97.9113
0 ± j 98.6960
.909396 ± j 127.683
0 ± j 128.304

Table 2 - EIGENVALUES OF THE CLOSED LOOP
SYSTEM ; ALL FEEDBACK LOOPS CLOSED.

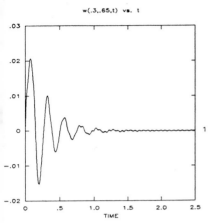

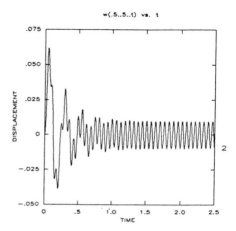

Fig. 2 - RESPONSE AT SEVERAL POINTS ON THE PLATE.

CONCLUSION

The response of a linear distributed parameter structure carrying R co-located sensors and actuators can be expressed as an infinite series in the modes of the closed loop system in which the feedback gains appear explicitly. This facilitates the solution of the related optimal control problem, as the displacement field and its temporal derivatives can be incorporated directly into an appropriate performance index and the gains found through a simple parameter optimization. When the system is controllable and observable and the performance index is quadratic in the displacement and velocity fields and control forces, the optimal gains can also be obtained from the solution of a matrix Lyapunov equation. These and various other aspects are presently being studied by the authors.

REFERENCES

1. Bergman, L. A. and Nicholson, J. W., "Forced Vibration of a Damped Combined Linear System," *ASME Journal of Vibration, Acoustics, Stress, and Reliability in Design*, Vol. 107, 1985, pp. 275–281.

2. Nicholson, J. W. and Bergman, L. A., "Free Vibration of Combined Dynamical Systems," *ASCE Journal of Engineering Mechanics*, Vol. 112, No. 1, 1986, pp. 1–13.

3. Bergman, L. A. and McFarland, D. M., "On the Vibration of a Point Supported Linear Distributed System," *ASME Journal of Vibration, Acoustics, Stress, and Reliability in Design*, Vol. 110, 1988, pp. 485–492.

4. McFarland, D. M. and Bergman, L. A., "Analysis of Passive and Active Discrete-Distributed Linear Dynamical Systems Using Green's Function Methods," Report UILU ENG 90-0504, University of Illinois at Urbana-Champaign, 1990.

U.S. National Workshop on Structural Control Research
Extended Summary - Implementation Techniques/Educational
Training Issues

Structural Control Research for Civil Systems-
Management Systems

Michael A. Cassaro, Knowlton Johnson
Center for Hazards Research and Policy Development
University of Louisville

The problems associated with renewing the nation's
infrastructure to be more resistant to the effects of
natural hazards are complex. Major problems affecting the
practice of routinely controlling the integrity of
structural resistance to hazards are:

o   widely separated institutional structures in which
     institutional systems are operated,

o   widely divergent values based on economics,
     politics, social issues, and culture,

o   ineffective lines of communication and access to
     information,

o   wide differences in levels of awareness, know-
     ledge, and understanding of the probabilities of
     loss and the consequences,

o   lack of an organized approach to education of the
     structural control process and development of the
     policy implementation requirements for planning,
     design, and construction to achieve natural
     hazards mitigation.

     This paper describes one solution to this problem by
linking institutional structures through a common network
for system risk assessment using a common database and
expert system for analysis. The process describes the
methods used to guide formation of policy for infrastructure
renewal, including repair, replacement, and new construction. The expanded management role of policy development for structure control considers maintenance procedures
coupled with recording practices and analysis leading to improvement in reliability of structural performance as a
systematic result through a management policy-development
initiative.

     The driving force of the management policy-development
initiative for structural control is a computerized risk-
based management model that utilizes system-specific inform-

ation and known parameters of component decay produced so that system managers can simulate stability and failure in their particular systems and thereby establish policy alternatives in the light of that information combined with cost alternatives. The process is completed through a technology transfer strategy that adapts the structure control model innovation within the user organization to routine use and application.

The use of change strategies, face-to-face communication activities, personal attention given to decision-makers throughout the implementation stage, and involvement of decision-makers in the implementation process leading to development of policy based on practical issues will be applied to a training program to assure technology transfer and implementation. The factors essential to predicting organizational readiness for change will be applied. These factors are:

o   ability of the users to apply the process will be assured,

o   values of the factors will be considered in the development of the training program,

o   adequate information will be provided to assure ease of application and understanding,

o   incentives for use of the models will be assisted through simplifying decision-making and achieving early results,

o   timeliness of activities will maintain a steady pace of learning and constant progress,

o   values to the organization and ready response to resistance to behavioral change will be dealt with in a positive fashion,

o   a product will be achieved and recognized with each effort,

o   the actions will be performance driven.

These factors are all to be applied in the technology transfer and implementation stage.

Implementation techniques require a concentrated training phrase that considers the values of the target group or audience (for example, a school board for a school system, or water company managers for a water distribution system, etc.), the identification of need, identification of deficiencies or obstacles to current decision-making performance, definition of training needs, and consideration of the organization environment.

The implementation of the process will be achieved through the technology transfer process which will assure the development of a routine posture of applying the policy and decision-making process for natural hazard mitigation within the organization. The changes in behavior required of the organization and its management call for a significant effort to secure complete cooperation, competent information exchange, adequate participation, and willingness to apply the strategies in the process.

Therefore, the training module should be carried out in cooperation with local or regional universities. The consortium of universities should be coordinated through regional organizations, such as ORAU or CUSEC, etc. The purpose of using regional or local universities is to encourage and foster research in technology transfer and establish a broad base for implementing techniques for natural hazard mitigation in which the training medium will apply complementary values and understanding of the organization in the technology transfer process. Applying control through regional organizations, such as ORAU/CUSEC will insure the uniformity of standards to the training modules and will enable the establishment of a mechanism for evaluation of the progress and success of the change strategies and implementation practice.

The above described model is a component of the presently active NSF project entitled, "A Prototype Risk Management Program for Water Companies in Earthquake Regions," conducted by the Center for Hazards Research and Policy Development of the University of Louisville.

IDENTIFICATION OF STRUCTURAL SYSTEMS USING NEURAL NETWORKS

by

Anastassios G. Chassiakos
School of Engineering
California State University
Long Beach, California 90840

Sami F. Masri
Department of Civil Engineering
University of Southern California
Los Angeles, California 90089-2531

Summary

This study introduces a new method for the identification of structural systems based on the use of neural networks. It is demonstrated that neural nets can be used effectively for the identification of linear as well as nonlinear dynamical systems.

Although several well established techniques exist for identifying linear time invariant systems, new methods are needed for the identification of nonlinear structural systems. This work uses an important class of neural networks, namely multilayer feedforward networks, for indentification purposes.

The neural net is given pairs of input/output "experimental" data. These pairs typically consist of "measured" (or simulated) excitation history and "measured" (or simulated) system responses. The network is then "trained" using these input/output pairs. The objective of training is to "teach" the network to produce a desired output given a corresponding input. During training, the network adjusts its own parameters in such a way as to minimize some error criterion between desired output and actual network response. The method by by which this adjustment is done is termed the "learning algorithm." In the present work, we focus on the "back propagation" learning algorithm and some modification of it.

First the indetification of linear time invariant structural systems is investigated. The simulated system responses to sinusoidal and random excitation are presented the neural net. The back propagation algorithm minimizes a square error criterion by adjusting the network's parameters. Once all the parameters or "weights" have been adjusted, then the network is given other excitations and its responses are compared to those of the original linear system under the same excitations. The results are extremely good and there is an almost perfect match, meaning that the neural network approach is very effective for the identification of linear time invariant structural systems.

The level of complexity is increased by considering systems with polynomial nonlinearities in the restoring force terms. After training, the network is able to identify the unknown system parameters and it gives responses very close to the desired outputs.

Other forms of nonlinearities as well as other forms of structural systems are currently under investigation. Several different network topologies and learning algorithms are being developed and tested for the identifiction of different classes of nonlinear systems. The results obtained so far are very encouraging. It is believed that the neural network approach will become more important in the future, particularly for on-line identification, as more computing power becomes available and with advances in parallel processing and hardware implementations of neural networks.

STRUCTURAL CONTROL RESEARCH AT NATIONAL CENTER FOR EARTHQUAKE ENGINEERING RESEARCH OF THE REPUBLIC OF CHINA

H.T. Chen
Head, Fundamental Research Division, NCEER–ROC, R.O.C.
C.H. Loh
Professor, Department of Civil Engineering, National Taiwan University, R.O.C.
D.S. Juang
Associate Professor, Department of Civil Engineering, National Central University, R.O.C.

Introduction

The National Center for Earthquake Engineering Research of the Republic of China (NCEER–ROC) was established in March 1990 by the National Science Council of the Republic of China. The structural control research has been identified as one of the research needs by the center. This paper will describe the organization of the NCEER–ROC and the proposed topics for the future structural control research.

Organization of National Center for Earthquake Engineering Research of The Republic of China

Taiwan is located in one of the most active seismic regions in the world and the earthquakes occur frequently. Over the past decades the Republic of China has enjoyed tremendous economic growth and many large construction projects and high–rise buildings have been planned or constructed. To avoid the destruction of these structures during the earthquakes, it is important that they be designed to resist the earthquakes. On November 15, 1986, a moderate earthquake struck Taiwan area and caused casualties, damages to several high–rise buildings and the collapse of a market building. It was then considered that more efforts should be directed to promote the earthquake engineering research and to upgrade the quality of earthquake–resistant design in Taiwan.
Through a series of discussions and evaluations, the National Science Council of the Republic of China finally decided to establish a National Center for Earthquake Engineering Research which is officially set into operation in March 1990 and is headquartered on the campus of National Taiwan University in Taipei. Its main objectives are as follows.
1. Promote the earthquake engineering research to mitigate the hazards due to earthquakes.
2. Integrate the research direction and coordinate a team effort in the study of earthquake engineering.
3. Facilitate information exchange among researchers and practitioners from different disciplines and the transfer of research results for industrial application to upgrade the quality of earthquake–resistant design.
4. Provide excellent environment to educate the talent in the field of earthquake engineering.
Figure 1 shows the organization of the NCEER–ROC. The Center is under the auspices of a Supervisory Committee and the function of the advisory Committee is to advise the director on the matters with regard to the center's research direction and operation and to review the research projects carried out by the center. Under the director, there are six divisions. The Fundamental Research Division and the Applied Research Division are to plan and to coordinate the basic research and applied research in the earthquake engineering, respectively. The Information Service Division is to provide earthquake–related research and reference assistance to the researchers, practicing engineers, and general public and to facilitate the information exchange and technology transfer through continuous education, workshops and conferences. In addition to planning

and coordinating the researches using the shaking table and pseudodynamic test system, the Earthquake Simulation Laboratory Division is responsible for the construction of a laboratory which will house a 5Mx5M tri–axial shaking table and pseudodynamic test system and their maintenance and management. The Field Experiment Park Division will manage a field experiment park with an area of approximately $2100M^2$ located in I–Lan where seismicity is active and plan and coordinate the construction of different types of model to study the responses of structures during the real earthquakes.

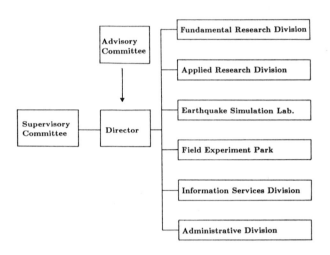

Figure 1: Organization of NCEER–ROC

Future Research on Structural Control at NCEER–ROC

The structural control research has been identified as one of the research needs by the NCEER–ROC. A panel on the Structural Control and Identification will be formed and its function is to develop a plan for the research in this field and to identify and to prioritize future research topics taking into account the resources available at the center and in Taiwan. Since the NCEER–ROC has just started its operation, many research topics are still under discussion. The researches on the active and passive control done over the past years in Taiwan were mainly analytical; the earthquake simulation facilities and the field experiment park of the NCEER–ROC now can provide better environment for the experimental study. As a result, the future structural control research will put emphasis on both the analytical and the experimental studies and the preliminarily proposed topics for the future structural control research are as follows.

1. Passive Control (Base–Isolation) System

Analytical studies will be performed to investigate the following topics: (1) long period effects, (2) effect of randomness of parameters, (3) interaction effects between horizontal seismic motions and vertical torsional and rocking modes, and (4) design procedures. For the experimental study, the following tests will be conducted: (1) validation test, (2) large scale model test and (3) material properties tests.

2. Active Control System

In the active control system the analytical studies will be made for the following topics: (1) optimal location of sensors, (2) time delay in control system, (3) control algorithm, (4) randomness in system parameters and (5) reliability issues. For the experimental investigations the emphasis will be placed on (1) full–scale model test, (2) aging effect, (3) reliability study and (4) development of efficient computer code.

Summary

In this paper the organization of the newly established National Center for Earthquake Engineering Research of the Republic of China is briefly described. A preliminary list of future research topics in the field of structural control for the center is also presented. It is the preliminary proposal and may be subjected to further modifications when more inputs are available. The list may not be extensive, but it indicates the determination of NCEER–ROC to promote further researches in the field of structural control.

USE OF A VARIABLE DAMPER FOR HYBRID CONTROL
OF BRIDGE RESPONSE UNDER EARTHQUAKE

Q. Feng[1] and M. Shinozuka[2]

ABSTRACT

This study demonstrates, by way of numerical simulation, the potential utility of variable dampers for controlling structural response under seismic loads. Standard control theory is difficult to apply in a straightforward fashion in this case where the equation of motion includes a nonlinear term caused by the damping coefficient that is a function of state variables. The present study deals with continuous girder bridges and examines the effect of bang-bang control and instantaneous optimal control on the reduction of absolute acceleration of the girder and relative displacement between the girder and support.

1. INTRODUCTION

As high-tech devices become more readily available, and their functional reliability more enhanced, they are beginning to find application in civil engineering. One of the more celebrated applications is the use of active control devices to suppress the vibrational response of large-scale structures such as high-rise buildings subjected to wind or earthquake loading. In theory, these active control systems have potential to perform better than passive control systems. However, one of the problems associated with these active control systems is that large active control forces and expensive control devices are needed because civil engineering structures are usually massive and heavy.

Recently, variable dampers are beginning to be used in automotive and other related fields for the purpose of control of mechanical vibration. Some of these are electro-rheological dampers whose damping coefficients can be controlled electrically, requiring much less external energy than actuator-based control devices would. For this reason, use of variable dampers provide a viable option for the control of civil engineering structures. The present study intends to explore the potential utility of variable dampers primarily for the control of bridge vibration under earthquake ground motion.

In passive isolation systems, relative displacement between the bridge girder and piers tends to be excessive. Increasing damping of the isolation system can reduce relative displacement, but will decrease the isolation effectiveness due to the corresponding increase

---

1. Research Assistant, Department of Civil Engineering and Operations Research, Princeton University, Princeton, NJ
2. Visiting Capen Professor of Structural Engineering, University at Buffalo, State University of New York, NY; On leave from Princeton University

in absolute acceleration. To alleviate this situation, therefore, the use of a hybrid control system is contemplated. The system is idealized as a model consisting of an elastic spring coupled in parallel with a damper having a controllable viscous damping coefficient. Conceptually, the coefficient of linear viscous damping can be thought of as consisting of two parts; a constant part c_0 and a variable part $c(t)$ in such a way that $c_0+c(t) \geq 0$, although $c_0+c(t)$ represents the variable damping coefficient of a physical damper.

The time-varying damping coefficient makes it rather difficult to apply more standard control theory. In view of this, two control algorithms, bang-bang control and instantaneous optimal control are proposed. The effectiveness of these control algorithms is examined by simulation. The results show some promise in that the hybrid isolation system with variable dampers is useful in reducing seismic responses of bridges.

2. CONTROL ALGORITHMS AND THEIR EFFECTIVENESS

2.1. Analytical Model

A continuous girder bridge is considered for analysis. The hybrid isolation devices are installed between the girder and piers. considering the piers to be rigid, the bridge can be modeled as a rigid mass with a single degree of freedom as shown in Fig.1.

The equation of motion of the system in the longitudinal direction under earthquake excitation is expressed as

$$m\ddot{x}(t)+kx(t)+c_0\dot{x}(t)+c(t)\dot{x}(t) = -m\ddot{z}(t) \tag{1}$$

where

    m : mass of bridge girder
    k : stiffness of isolation system
    c_0 : constant damping coefficient of variable damper in isolation system
    $c(t)$: variable damping coefficient of variable damper in isolation system
    $x(t)$: relative displacement of girder to pier
    $\ddot{z}(t)$: input earthquake acceleration

The values of parameters used for simulation are listed below (Ref.1 and 2).
    m = $1.020*10^6$ kg
    ω = 1.795   ($T = 3.5$ sec)   ($\omega = \sqrt{k/m}$)
    ζ_0 = 0.1   ($2\omega\zeta_0 = c_0/m$)

2.2. Bang-bang Control

In a passive isolation system, the damping has opposite effects on the absolute acceleration of the girder and on the relative displacement between the girder and piers. It is anticipated that by controlling the variable damping coefficient, both absolute acceleration and relative displacement can be reduced.

For this purpose, nondimensional absolute acceleration and relative displacement are defined as $|\ddot{x}_a|/\ddot{x}_{aref}$ and $|x|/x_{ref}$ respectively by introducing reference values of acceleration $\ddot{x}_{aref}$ and displacement x_{ref}. As mentioned earlier, the isolation device produces opposite effects on the acceleration and displacement. Therefore, it is interpreted that an optimality has been achieved if the response follows along line OP in the nondimentional acceleration vs. displacement plane as shown in Fig.2 (Ref.3). The response along OP in fact implies that the variation of the acceleration relative to its reference value is always the same as the variation of the displacement relative to its reference value. Therefore, when the nondimensional relative displacement is larger than the nondimentional absolute acceleration (domain R in Fig.3), the damping coefficient should be increased. When the nondimensional absolute acceleration is larger than the the nondimensional relative displacement (domain Q) , the damping coefficient should be decreased. On the basis of this observation, the following bang-bang control algorithm is proposed.

$$c(t) = \begin{cases} c_{max} & \text{if response is in R} \\ c_{min} & \text{if response is in Q} \end{cases} \qquad (2)$$

The control system is shown in Fig.3. Signals $|\ddot{x}_a|$ and $|x|$, detected by an acceleration sensor installed in the girder and by a displacement sensor installed between the girder and pier respectively, are sent to a controller (microcomputer). The computer switches the value of the damping coefficient according to the control law in Eq.(2) to control the response acceleration and displacement. Since $c_0 + c(t) \geq 0$, the smallest algebraic value of $c(t)$ is $-c_0$ and therefore,

$$c_{min} \geq -c_0 \qquad (3)$$

2.3.. Instantaneous Optimal Control
The optimal damping coefficient can be determined by minimizing the following time dependent objective function J(t) at every time instant t for the entire duration of an earthquake (Ref.4).

$$J(t) = q_1 \ddot{x}_a^2(t) + q_2 x^2(t) + rc^2(t) = X(t)^T Q(t) X(t) + rc^2(t) \qquad (4)$$

where

$$X(t) = \begin{bmatrix} x(t) \\ \dot{x}(t) \end{bmatrix}, \qquad Q(t) = \begin{bmatrix} q_1 \dfrac{k^2}{m^2} + q_2 & q_1 \dfrac{k(c_0 + c(t))}{m^2} \\ q_1 \dfrac{k(c_0 + c(t))}{m^2} & q_1 \dfrac{(c_0 + c(t))^2}{m^2} \end{bmatrix}$$

Numerical solutions of the nonlinear equation of motion, Eq.(1), are obtained in Eq.(6) by Wilson's θ method, and are used as constraints when minimizing J(t).

$$X(t) = D(t - \Delta t) + A_1 \ddot{z}(t) + A_2 c(t) \dot{x}(t) \qquad (5)$$

where

$$D(t - \Delta t) = A_3 X(t - \Delta t) + A_4 c(t - \Delta t) \dot{x}(t - \Delta t) + A_5 \ddot{z}(t - \Delta t)$$

with A_1, A_2, A_3, A_4 and A_5 being determined by k, m, c_0 and control time interval Δt. Thus, the generalized objective function is established by introducing a Lagrangian multiplier vector $\lambda(t)$ to the objective function:

p 109

$$H = X(t)^T Q(t) X(t) + rc^2(t) + \lambda^T(t)\{X(t) - D(t-\Delta t) - A_1 \ddot{z}(t) - A_2 c(t) \dot{x}(t)\} \quad (6)$$

The necessary conditions for minimizing the objective function $J(t)$ are:

$$\frac{\partial H}{\partial X} = 0, \quad \frac{\partial H}{\partial c} = 0, \quad \frac{\partial H}{\partial \lambda} = 0 \quad (7)$$

Substitution of Eq.(6) into Eq.(7) leads to

$$2Q(t)X(t) + [I - A_2[0 \quad c(t)]]^T \lambda(t) = 0 \quad (8)$$

$$2rc(t) - A_2^T \lambda(t)\dot{x}(t) + 2q_1\frac{(c_0+c(t))}{m}\dot{x}(t) + \frac{k}{m}x(t)\frac{\dot{x}(t)}{m} = 0 \quad (9)$$

$$[I - A_2[0 \quad c(t)]]X(t) - D(t-\Delta t) - A_1 \ddot{z}(t) = 0 \quad (10)$$

Thus, the optimal damping coefficient $c(t)$ and state vector $X(t)$ at each time instant t can be obtained from Eqs.(8)-(10).

2.4. Simulation Results

In order to examine the effectiveness of the bang-bang control and instantaneous optimal control algorithms proposed above, simulation analysis using Eq.(1), with the parameter values assumed in Section 2.1 has been performed. The earthquake record EL-CENTRO (NS,1940) linearly adjusted to peak acceleration of 200 gal was used. In numerical simulations, the variability of the dampers is restricted to the range:

$$c_{min} = -c_0 \leq c(t) \leq 3c_0 = c_{max} \quad (11)$$

Figure 4 shows the simulation results, where the response absolute accelerations of the bridge girder that is not isolated, passively isolated and isolated by the present hybrid system are compared in terms of their respective peak values. Similarly, relative displacements are also compared. Figure 5 shows, for comparison, response time histories of the bridge girder that is not isolated, passively isolated and isolated by hybrid system controlled with the instantaneous optimal algorithm. The same figure also shows time histories of the input ground motion (which equals the absolute acceleration of the girder if the girder is not isolated and if the piers are considered rigid.) and the variation of damping coefficient. For the girder which is not isolated, the natural period and damping ratio of the bridge are assumed to be $T_b = 0.5$ sec and $\zeta_b = 0.02$, respectively.

These results show that passive isolation supports greatly reduce the response acceleration of the bridge girder, but at the same time produced a large relative displacement between the girder and the piers. However, by controlling the variable damping coefficient using proposed control algorithms, the displacement can be further decreased significantly, and at the same time, the acceleration can also be decreased although by a relatively small amount below the low level already achieved by the passive control system. To be precise, the peak relative displacement can be decreased by 41% and peak absolute acceleration by 22% with the aid of instantaneous optimal control, compared with corresponding values under passive isolation.

3. CONCLUDING REMARKS

A hybrid isolation system using variable dampers for bridges is investigated. The effectiveness of controlling damping coefficients by the proposed control algorithms is demonstrated by numerical simulation.

The following conclusions are obtained from the present study.

(1) Compared with the passive isolation system, the hybrid isolation system using variable dampers exhibits superior isolation performance both in the reduction of response absolute acceleration and relative displacement.

(2) Both bang-bang control and instantaneous optimal control algorithms are effective. The bang-bang control is easier to implement in the actual system because the algorithm is extremely simple, although the effectiveness of this control is slightly inferior to the instantaneous optimal control.

ACKNOWLEDGEMENT

This work was partially supported by the National Center for Earthquake Engineering Research under Grant Number 90-2204. The first author wishes to acknowledge the advice and encouragement provided by Professor T. Fujita of the University of Tokyo. She is also appreciative of Daikin Industries Limited, Japan for its support on her behalf through Princeton University. The authors are also grateful to Dr. K. Kawashima of the Public Works Research Institute, Ministry of Construction, Japan for his valuable comments on the subject matter.

REFERENCES

1. Kawashima, K. et al., "Shaking Table Test on Dynamic Response of Base-isolated Bridges", Civil Engineering Journal, Vol. 30-10, (1988)
2. Kawashima, K. et al., "Active Control of Seismic Response of Structure by Means of Mass Damper", Civil Engineering Journal, Vol. 31-5, (1989)
3. Tsujiuchi, N., Koizumi, T., Kishimoto, F. and Fujita, T., "Control of Structural Vibration by Isolation Device Using Adaptive Control", Proc. of JSME, No.900-4,Vol.B, (1990)
4. Yang, J.N., et al., "Optimal Control of Nonlinear Flexible Structures", Technical Report NCEER-88-0002, (1988)

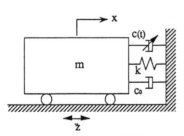

Fig.1 Analytical Model

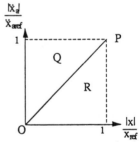

Fig.2 Optimal Performance of Isolation System

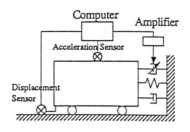

Fig. 3   Diagram of
Bang-bang Control System

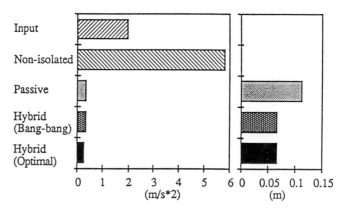

Fig.4   Performance of Hybrid Isolation System (Peak Values)

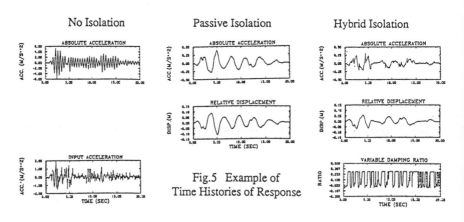

Fig.5   Example of
Time Histories of Response

RESEARCH ON VIBRATION CONTROL OF STRUCTURES AT THE UNIVERSITY OF ILLINOIS

D. A. Foutch[1] and J. Ghaboussi[1]

Several research projects on vibration control of structures are currently underway or are in the planning stages at the University of Illinois at Urbana-Champaign. Vibration control is interpreted in its broadest sense, which includes both active and passive systems. This report is intended to provide a brief description of this research.

Research on Active Vibration Control of Structures

The structural control research by J. Ghaboussi and his co-workers is concentrated in two areas: 1) experimental research in digital vibration control of structures; and 2) use of neural networks in vibration control of structures. Other related areas addressed by this research group include the use of active control methodology, along with the state-of-the-art sensor technology and innovative actuator designs, to conduct concurrent analysis and testing of complex engineering problems. In the next paragraphs, our research in the two areas of vibration control will be briefly described.

Experimental Studies in Digital Vibration Control

The purpose of this research is to study the fundamental problems of vibration control of structures when the control is done with an on-line computer and to develop methodology for improving the effectiveness of digital vibration control. The experimental part of this research was carried out on a simple two degree-of-freedom system consisting of masses attached to actuators, with the top mass and actuator system mounted on the bottom mass. The bottom mass was moved with a prescribed motion, while the top actuator was used to control the motion of the top mass. This experimental setup was specifically designed for the purpose of study of digital control problems and by itself it did not represent an actual structural system. The results of our research so far indicates the most effective control strategies may involve a combined active and passive control system. Currently, we are studying various methods of integrating passive control devices in the control loop.

---

[1]Professor, Department of Civil Engineering, University of Illinois at Urbana-Champaign, 205 North Mathews, Urbana, IL 61801

Structural Vibration Control with Neural Networks

In design and implementation of any structural control scheme for large and complex structural systems, two important steps are identification and mathematical modeling. Even the most advanced numerical models of the complex structural systems may not be adequate for the purpose of control. In addition to discretization errors, such numerical models introduce time delays and may cause spillover from the unmodeled dynamics of the system. Our approach is to use the learning algorithms in artificial intelligence to replace the numerical model of the structure and the control algorithm using that numerical model. The basic strategy in this research is to use neural networks in the control loop. Initially, the neural network receives sensor data and "learns" about the structural behavior. This knowledge is then stored in the connection strengths of the neural network. The trained neural network is then used to control the structure. The resulting control system is fault tolerant and adaptive. When the structural behavior changes during the control process, the neural network continues to adapt and learn about the structural behavior, and in the process it modifies the strengths of its connections to reflect the new knowledge of the structural behavior.

Research on Passive Vibration Control of Structures

Research on passive control of structures is being planned by S. L. Wood and D. A. Foutch in cooperation with researchers from the U.S. Army Construction Engineering Research Lab located in Champaign, Illinois. The purpose of the research is to investigate the vulnerabilities of nonductile reinforced concrete frames characteristic of DOD facilities and to determine effective retrofit techniques to improve the behavior and survivability of these structures under severe earthquake loadings.

One retrofit technique that is of particular interest is the incorporation of added damping to these structures. This technique has been shown through scale model tests to be very effective in reducing the earthquake response of steel frame buildings. One scheme that will be tested will involve dampers made from viscoelastic materials. Other materials will also be investigated.

One difficulty that is faced when applying this retrofit method to nonductile RC buildings is that these structures are stiffer than steel frames, and they exhibit brittle behavior. Therefore, the tolerable interstory displacements are probably smaller than they would be for a steel frame.

The research on passive control will be conducted in two phases. In Phase 1, the feasibility of using added damping will be investigated. Half-scale SDOF models of a single joint area with column sections will be tested on the earthquake simulator at CERL. These will be reinforced in a similar fashion to existing nonductile concrete frames. Some specimens will be tested with dampers and some without. In Phase 2, model building frames of approximately one-fourth scale will be tested with and without dampers. Through this program, it is expected that the effectiveness and efficiency of various retrofit schemes will be determined.

EFFICACY OF ACTIVE CONTROL IN STRUCTURAL ENGINEERING

Dan M. Frangopol
Professor Civil Engineering
Department of Civil Engineering
University of Colorado at Boulder
Boulder, CO 80309-0428

The ultimate goal in structural design is to find the optimum balance between conflicting requirements such as safety and cost. Historically, the desire to evaluate the safety of structural systems has been a strong driving force behind the development of modern structural system reliability methods. Today, the need for using system reliability methods into both design and evaluation of structural systems is evident from recent failures in buildings, bridges, offshore platforms and other structures. In most instances, failure investigations revealed the lack of structural system redundancy. The protection of a structural system under extraordinary loads may be given by both alternative load paths due to built-in redundancy and by extra redundancy provided by active structural control.

Research is needed to develop a framework for consideration of efficacy of active control in structural engineering. Proposed measures should be based on structural system reliability levels with and without active control.

PHASE–ADJUSTED ACTIVE CONTROL OF STRUCTURES WITH IDENTIFICATION OF RANDOM EARTHQUAKE GROUND MOTION

Hirokazu IEMURA* , Yoshikazu YAMADA** , Kazuyuki IZUNO***
Yoshihisa IWASAKI**** and Satoru OHNO*****

ABSTRACT

Phase-adjusted active control of structures under stochastic base excitation is investigated. The synthesis of an active control system discussed in this study is based on stochastic control theory in which an energy-based stochastic criterion is employed for optimal control. Accordingly, an optimal control law for input with a predominant frequency is developed which consists of a response feedback portion and a ground-input feedforward portion even when the dynamic properties of the ground are not known. The effectiveness of the phase-adjusted active control law including identification of ground motion is demontrated with a numerical simulation, which shows that the proposed control law brings considerable control force (RMS) reduction compared with the feedback control for the same level of response reduction in both cases.

INTRODUCTION

With vast improvement in recent construction techniques and materials, light-weight and flexible structures such as high-rise buildings and long-span bridges have been designed and constructed in Japan. It is becoming critically important to suppress dynamic response of structures due to wind and earthquake excitations, not only for their safety but also their serviceability.

First, very high dynamic amplification of structural response due to the narrow frequency banded ground motions with long duration observed at the soft soil site in recent earthquakes is discussed. Introduction of the active control force to suppress the resonant phenomena of flexible and lightly damped structures is one of approaches to get high reliability of structural systems.

Second, active control of structures under stochastic base excitation is investigated. The synthesis of an active control system discussed in this study is based on the use of stochastic control theory, where an energy-based stochastic criterion is employed for optimal control. As a result, the optimal control law for input with a predominant frequency is developed, which consists of a response feedback portion and a ground-input feedforward portion.

Third, to apply above theory even when the ground dynamic properties are not known, identification technique using extended Kalman filtering algorithm is introduced.

Fourth, to overcome the effects of phase lag between measurement and activation of control force, earthquake ground motion and structural response are predicted using Kalman filtering technique. Feedforward gain is multiplied to the predicted ground motion and feedback gain is multiplied to the predicted structural response to obtain the optimal control force.

*   Associate Professor, Department of Civil Engineering, Kyoto University
**   Professor, Department of Civil Engineering, Kyoto University
***   Research Associate, Department of Civil Engineering, Kyoto University
****   Engineer, Kobe City Municipal Government
***** Engineer, Ohbayashi Cooporation

The effectiveness of the phase-adjusted active control law including identification of ground dynamic properties is demonstrated with a numerical simulation, which shows that the proposed control law brings considerable control force reduction, compared with the conventional feedback control.

HIGH DYNAMIC AMPLIFICATION OF STRUCTURAL RESPONSE DUE TO RECENT EARTHQUAKES

In conventional earthquake engineering, earthquake ground acceleration has been considered as nonstationary random motion with broad band of frequency components. The El Centro Records of the 1940 Imperial Valley Earthquake is a typical case. This is still true for sites with good soil conditions. However, narrow frequency banded ground motion with long duration has been observed in the recent earthquakes at soft soil sites, where modern structures have been constructed. When subjected to this type of earthquake ground motion, flexible and lightly damped structures have shown resonant phenomena, which leads to high dynamic amplification of structural response. Examples of structural damages caused by the high amplification effects due to some recent earthquakes are described in the following;

Nihonkai-Chubu Earthquake in 1983 in JAPAN

Very high sloshing in oil tanks was observed at Niigata City which is 270 km away from the epicenter. Niigata is covered by deep and soft sand. Ground motion with period of 10 seconds became predominant and continued for more than several minutes. Mended and corrected displacement and acceleration-type seismograms were found to give much higher response spectra than the design values for long-period (5-10 sec) structures with 2 ~ 5% damping.

Mexico Earthquake in 1985

In Mexico City which is 350 km away from the epicenter, ground motion with predominant period of 2 seconds. Acceleration response spectra with 2% damping shown in Fig. 1 is much higher than the design spectra. The acceleration response at 2 seconds is approximately 10 times amplified compared to the ground acceleration, which caused significant damage to 15 ~ 20 storied buildings.

Armenia Earthquake in 1988 in USSR

Severe damage concentration in Leninakan which is 30 km away from the epicenter is considered to have correlations with local soil conditions. Leninakan lies in an alluvial valley with 200 ~ 300 m of underlyning sedimentary deposit. Aftershock measurement by U.S.G.S. verifies high dynamic amplification.

Loma Prieta Earthquake in 1989

Although the earthquake was not really big one, severe damages occured in soft soil sites. Building damage in Marina district and Viaduct damage in Cypress area are typical examples. Significant differences in acceleration records are found in hard and soft soil conditions.

In designing flexible and lightly damped structures, we have to accept high dynamic amplification factor of structural response, which leads to high sesmic design force. An alternative design approach is to introduce active control force to suppress the resonant effect of structural response.

OPTIMAL CONTROL UNDER RANDOM EXCITATION
WITH A PREDOMINANT FREQUENCY

An equation of motion of n degrees-of-freedom (DOF) structures subjected to earthquake ground displacement $z(t)$ with m-vector control force $u(t)$ is written as

$$M\ddot{y} + C\dot{y} + Ky = -\underline{m}\ddot{z} + Du \tag{1}$$

where, M, C, K are the mass, damping and stiffness matrices of $n \times n$ dimensions, y is n-vector relative displacement response, $\underline{m}$ is n-vector mass, and D is a $n \times m$ matrix. The above equation of motion can be described by the following state differential equation:

$$\dot{x} = Ax + Bu + G\ddot{z} \tag{2}$$

where,

$$x = \begin{pmatrix} y \\ \dot{y} \end{pmatrix} \qquad A = \begin{pmatrix} o & I \\ -M^{-1}K & -M^{-1}C \end{pmatrix} \qquad B = \begin{pmatrix} 0 \\ M^{-1}D \end{pmatrix} \qquad G = \begin{pmatrix} 0 \\ -M^{-1}\underline{m} \end{pmatrix}$$

In earthquake response spectra, it is often found that ground motion is affected by soft surface layer. It is reasonable to assume that surface ground motion is the dynamic response of surface layer subjected to random excitation of base rock, from which the Kanai-Tajimi model is developed. Letting Z denote the state vector of the ground, an equation of motion is written as

$$\dot{Z} = A_z Z + G_z w \tag{3}$$

where, w : a white-noise process with intensity V_0

$$Z^T = (z, \dot{z}) \qquad A_z = \begin{pmatrix} 0 & 1 \\ -\omega_g^2 & -2\zeta_g\omega_g \end{pmatrix} \qquad G_z^T = (0, 1)$$

Then, the power spectrum of the ground acceleration is obtained as

$$P(\omega) = |H(i\omega)|^2 = \frac{\left(\dfrac{\omega}{\omega_g}\right)^4}{\left\{\left(\dfrac{\omega}{\omega_g}\right)^2 - 1\right\}^2 + 4\zeta_g^2\left(\dfrac{\omega}{\omega_g}\right)^2} V_0 \tag{4}$$

Combining Eq. (2) for structures and (3) for surface ground layers, an equation of motion of total system is described as

$$\dot{\tilde{x}} = \tilde{A}\tilde{x} + \tilde{B}u + \tilde{G}w(t) \tag{5}$$

where,

$$\tilde{x} = \begin{pmatrix} x \\ Z \end{pmatrix}_{(n+2)\times 1} \qquad \tilde{A} = \begin{pmatrix} A & GD_z \\ O & A_z \end{pmatrix}_{(n+2)\times(n+2)}$$

$$\tilde{B} = \begin{pmatrix} B \\ O \end{pmatrix}_{(n+2) \times m} \qquad \tilde{G} = \begin{pmatrix} G \\ G_z \end{pmatrix}_{(n+2) \times 1} \qquad D_z = \begin{pmatrix} 0 & 1 \end{pmatrix} A_z$$

As an optimal control strategy, minimization of the objective function with quadratic form is employed. Assuming that $\ddot{z}(t)$ and $y(t)$ are stationary random processes, the expected total energy of the structural system is expressed as,

$$J = E\left[\tilde{x}^T \tilde{R}_1 \tilde{x} + u^T R_2 u \right] \longrightarrow \min \tag{6}$$

With the Riccati matrix $\tilde{P}$ calculated from next equation,

$$\tilde{R}_1 - \tilde{P}\tilde{B}R_2^{-1}\tilde{B}^T\tilde{P} + \tilde{A}^T\tilde{P} + \tilde{P}\tilde{A} = O \tag{7}$$

the optimal control force $u(t)$ is obtained as

$$u(t) = -R_2^{-1}\tilde{B}^T\tilde{P}\tilde{x}(t) = \tilde{F}\tilde{x}(t) = F_x x + F_z Z \tag{8}$$

From the above equation, it is easily understood that the optimal control force is a function of the structural response x and the input ground motion Z. This is schematically illustrated in Fig. 2, where $F_x x$ and $F_z Z$ are called closed-loop and open-loop control, respectively. F_x and F_z are called feedback and feedforward gain, respectively.

IDENTIFICATION OF GROUND PARAMETERS

In using Eq. (3), we have to know the natural frequency of the ground ω_g and the damping constant ζ_g. To get the optimal feedforward gain even when these parameters are not known, identification theory is introduced.

Using the linear acceleration method, Eq. (3) can be expressed in the discretized form;

$$z_{t+1} = A_z z_t + G_z w_{t+1} \tag{9}$$

where,

$$z_t = \begin{pmatrix} z_t \\ \dot{z}_t \\ \ddot{z}_t \\ \zeta_{gt} \\ \omega_{gt} \end{pmatrix} \qquad G_z = \frac{1}{1 + \Delta t \zeta_g \omega_g + \frac{\Delta t^2}{6}\omega_g^2} \begin{pmatrix} \frac{\Delta t^2}{6} \\ \frac{\Delta t}{2} \\ 1 \\ 0 \\ 0 \end{pmatrix} \qquad \Delta t: \text{interval of calculation}$$

$$A_z = \frac{1}{1 + \Delta t \zeta_g \omega_g + \frac{\Delta t^2}{6}\omega_g^2} \begin{pmatrix} 1 + \Delta t \zeta_g \omega_g & \Delta t + \frac{2}{3}\Delta t^2 \zeta_g \omega_g & \frac{\Delta t^2}{3} + \frac{\Delta t^3}{6}\zeta_g \omega_g & 0 & 0 \\ -\frac{\Delta t}{2}\omega_g^2 & 1 - \frac{\Delta t^2}{3}\omega_g^2 & \frac{\Delta t}{2} - \frac{\Delta t^3}{12}\omega_g^2 & 0 & 0 \\ -\omega_g^2 & -2\zeta_g \omega_g - \Delta t \omega_g^2 & -\Delta t \zeta_g \omega_g - \frac{\Delta t^2}{3}\omega_g^2 & 0 & 0 \\ 0 & 0 & 0 & 1 + \Delta t \zeta_g \omega_g + \frac{\Delta t^2}{6}\omega_g^2 & 0 \\ 0 & 0 & 0 & 0 & 1 + \Delta t \zeta_g \omega_g + \frac{\Delta t^2}{6}\omega_g^2 \end{pmatrix}$$

As the equation above is nonlinear for ω_g and ζ_g, extended Kalman filtering technique is used to identify these parameters. The filtering algorithm is shown in Table 1.

p 119

PHASE–ADJUSTED ACTIVE CONTROL

In order to overcome the effects of phase lag between measurement and activation of control force, earthquake response is predicted using Kalman filtering technique; and from the predicted response, optimal control force is applied.

Using the linear acceleration method, equation of motion of a SDOF structure with control force u is expressed in the following discretized form.

$$x_{t+1} = Ax_t + Bu_{t+1} + Gz_{t+1} \tag{10}$$

where,

$$x_t = \begin{pmatrix} x_t \\ \dot{x}_t \\ \ddot{x}_t \end{pmatrix} \qquad A = \frac{1}{m + \frac{\Delta t}{2}c + \frac{\Delta t^2}{6}k} \begin{pmatrix} m + \frac{\Delta t}{2}c & \Delta tm + 2\Delta t^2 c & \frac{\Delta t^2}{3}m + \frac{\Delta t^2}{12}c \\ -\frac{\Delta t}{2}k & m - \frac{\Delta t^2}{3}k & \frac{\Delta t}{2}m - \frac{\Delta t^2}{12}k \\ -k & -c - \Delta tk & -\frac{\Delta t}{2}c - \frac{\Delta t^2}{3}k \end{pmatrix}$$

$$B = \frac{1}{m + \frac{\Delta t}{2}c + \frac{\Delta t^2}{6}k} \begin{pmatrix} \frac{\Delta t^2}{6} \\ \frac{\Delta t}{2} \\ 1 \end{pmatrix} \qquad G = \frac{-m}{m + \frac{\Delta t}{2}c + \frac{\Delta t^2}{6}k} \begin{pmatrix} 0 & 0 & \frac{\Delta t^2}{6} \\ 0 & 0 & \frac{\Delta t}{2} \\ 0 & 0 & 1 \end{pmatrix} \qquad \Delta t : \text{interval of calculation}$$

However, phase lag between measurement and activation cannot be eliminated. In this study, x_{t+1} is predicted from x_t as follows.

$$\hat{x}_{t+1/t} = (I - BF)^{-1}\{Ax_t + (G + BF_z)\hat{z}_{t+1/t}\} \tag{11}$$

Hence optimal control force for the random excitation with a predominant frequency is obtained as

$$u_{t+1} = F_x\hat{x}_{t+1/t} + F_z\hat{z}_{t+1/t} \tag{12}$$

NUMERICAL EXAMPLES

As a model of flexible and lightly-damped structures, a single-DOF structure with damping ratio of 0.02 and natural period of 2 second ($m = 4\ kg \cdot sec^2/cm$) is adopted for numerical simulations.

From Fig. 3, it is found that prediction of ground motion and structural response in phase-adjusted control suppresses vibration energy of a structure effectively. If no prediction is performed, vibrational energy goes up very sharply with time delay. This is due to the predominance of frequency component in random excitation.

Time history displacement response of structures and control force is shown in Fig. 4, for the case of 0.4 seconds delay time. It is found that the response of the phase-adjusted controlled structure with prediction and the optimally controlled structure without time delay have little difference. Corresponding control forces are also found to be very similar. These results verifies that the phase-adjusted control with prediction of ground motion and structural response proposed in this study works satisfactorily.

Then, the proposed algorithm is applied using SCT record of 1985 Mexico earthquake, which has a predominant frequency of 0.5 Hz. Figure 5 shows the estimated ground parameters ζ_g (the broken line) and ω_g (the solid line) at each time step. These values converge to the desired values after 5 seconds. From Fig. 6, the maximum displacement response is 70.4 cm when the structure is uncontrolled; 18.1 cm (25.7% of the uncontrolled case) when it is controlled with feedback gain; and 9.92 cm (14.1% of the uncontrolled case) when it is controlled with both feedback and feedforward gain. Using this algorithm, the structure can be well controlled even if the predominant frequency of the ground motion changes during an earthquake.

CONCLUSIONS

Major conclusions from this study are:

(1) Very high dynamic amplification of structural response due to the narrow frequency banded ground motions with long duration observed at the soft soil site in recent earthnquakes is pointed out. Introduction of the active control force to suppress the resonant phenomena of flexible and lightly damped structures is a promising approach to get high reliability of structural systems.
(2) When the earthquake ground motion have a predominant frequency, future input ground motion can be predicted using Kalman filtering technique and then structural response is predicted. Feedforward gain is multiplied to the predicted ground motion and feedback gain is multiplied to the predicted response to obtain the optimal control force.
(3) Phase-lag effect is reduced with prediction of input and response, as illustrated by a numerical simulation. However, if phase lag becomes more than one-third of the natural period of structure, efficiency of control force decreases rapidly.
(4) The predominant frequency can be estimated during an earthquake using extended Kalman filtering technique. This active control algorithm with identification of ground parameters can be applicable when the predominant frequency changes during an earthquake.

ACKNOWLEDGEMENT

The authors would like to thank Mr. William Tanzo, graduate student at Kyoto University, for his help in writing and arranging this paper.

REFERENCES

1. H. Kwakernaak and R. Sivan, *Linear Optimal Control Systems,* Wiley Interscience, 1972.
2. J. N. Yang and M. J. Lin, "Optimal Critical-Mode Control of Building under Seismic Load," *Proc. ASCE,* Vol. 108 (EM6), pp. 1167–1185, 1982.
3. L. Meirovitch and L. M. Silverberg, "Control of Structures Subjected to Seismic Excitation," *Proc. ASCE,* Vol. 109 (EM2), pp. 604–618, 1983.
4. H. H. W. Leipholz and M. Abdel-Rohman, *Control of Structures,* Martinus Nijhoff Publishers, 1986.
5. T. Kobori, "Active Seismic Response Control," *Proc. 9th WCEE,* Vol. 8, pp. 435–446, 1988.
6. Y. Yamada, H. Iemura and A. Igarashi, "Active Control of Structures under Stochastic Base Excitation with a Predominant Frequency," *Proc. of 5th ICOSSAR,* San Francisco, 1989.

7. Y. Yamada, H. Iemura, A. Igarashi and Y. Iwasaki, "Phase-Delayed Active Control of Structures under Random Earthquake Motion," *Proc. 4th United States National Conference on Earthquake Engineerings*, Vol. 3, pp. 447-456, 1990.

Table 1 Filter variables and Kalman filtering algorithm.

Variable	Definition	Dimension
$\hat{x}(k/k)$	State estimate at t_k given y_k	$n \times 1$
$P(k/k)$	Covariance matrix of the error in $\hat{x}(k/k)$	$n \times n$
$\Phi(k+1/k)$	State transition matrix (from t_k to t_{k+1})	$n \times n$
$\Gamma(k+1/k)$	System noise coefficient matrix	$n \times r$
$Q(k)$	System noise covariance matrix	$r \times r$
$\hat{x}(k+1/k)$	State estimate at t_{k+1} given y_k	$n \times 1$
$P(k+1/k)$	Covariance matrix of the error in $\hat{x}(k+1/k)$	$n \times n$
$M(k+1)$	Measurement matrix	$m \times n$
$R(k+1)$	Measurement noise covariance matrix	$m \times m$
$K(k+1)$	Filter (Kalman) gain matrix at t_{k+1}	$n \times m$
y_{k+1}	Measurement (observation) at t_{k+1}	$m \times 1$
1.	Store the filter state $[\hat{x}(k/k),\ P(k/k)]$;	
2.	Compute the predicted state $\hat{x}(k+1/k) = \Phi(k+1/k) \cdot \hat{x}(k/k)$;	
3.	Compute the predicted error covariance matrix $P(k+1/k) = \Phi(k+1/k) \cdot P(k/k) \cdot \Phi^T(k+1/k) + \Gamma(k+1/k) \cdot Q(k) \cdot \Gamma^T(k+1/k)$;	
4.	Compute the filter gain matrix $K(k+1) = P(k+1/k) \cdot M^T(k+1)[M(k+1) \cdot P(k+1/k) \cdot M^T(k+1) + R(k+1)]^{-1}$;	
5.	Process the observarion y_{k+1} $\hat{x}(k+1/k+1) = \hat{x}(k+1/k) + K(k+1)[y_{k+1} - M(k+1) \cdot \hat{x}(k+1/k)]$;	
6.	Compute the new error covariance matrix by $P(k+1/k+1) = [I - K(k+1) \cdot M(k+1)]P(k+1/k)$;	
7.	Set $k = k+1$, and return to step 1.	

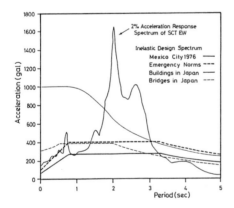

Fig. 1 2% Acceleration Response Spectra at SCT in Mexico City and Inelastic Design Spectra

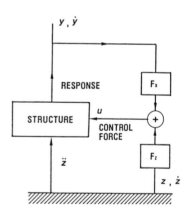

Fig. 2 Feedback and Feedforward Control System

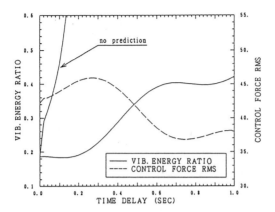

Fig. 3 Vibrational Energy of Phase-Delayed Controlled Structures
        With and Without Control Force Prediction
        (Filtered White-Noise Excitation, Feedback and Feedbackward Control)

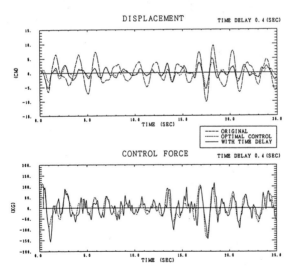

Fig. 4 Simulated Time History of Displacement Response and Control Force
        (Filtered White-Noise Excitation, Feedback and Feedbackward Control)

p 123

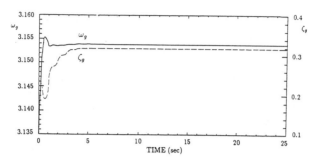

Fig. 5 Identified Dynamic Properties of Ground at Each Time Step

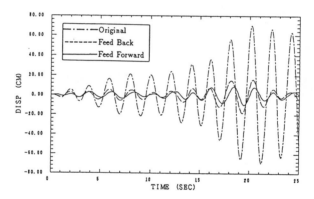

Fig. 6 Simulated Time History of Displacement Response
to SCT record of 1985 Mexico Earthquake

ACTIVE ISOLATION

Jose A. Inaudi * and James M. Kelly **

Abstract

The experimental implementations of two active isolation strategies are described in this paper. A single control force is applied on the first floor of a base isolated model. The goal of both controllers is to reduce the absolute motion and interstory drifts of the superstructure under seismic excitation. Two different approaches were tested, i) active damping and ii) state bounding. In the first approach, artificial damping is obtained by generating a control force as a non-linear function of the absolute velocity of the first floor (base). The second approach is a robust control scheme developed by A. Soldatos and G. Leitmann which could, in theory, make the absolute response of the structure arbitrarily small.

1. Active isolation

The concept of active isolation as an alternative approach in structural design for earthquake hazard mitigation is very recent [1,2]. Kelly, Leitmann and Soldatos introduced the concept of active isolation with the control objective of minimizing absolute displacements and velocities of the structure. Active control schemes designed for minimizing *absolute motion* of base isolated structures under seismic excitation introduce the possibility of defining the control force as function of the state of the system described in *absolute coordinates*. This resistant scheme is not obtainable by passive means and is very powerful for absolute motion reduction. The introduction of passive energy dissipation devices in base isolated structures is not as efficient as increasing the active damping. In order to illustrate this simple concept let's compare the amplitude X of the absolute response of a SDOF system with natural frequency ω under harmonic ground motion $g(t) = G \cos(\bar{\omega}t)$. The introduction of passive energy dissipation devices is modeled by a resistant force proportional to the relative velocity while the active damping is modeled as a force proportional to the absolute velocity of the mass. The magnification factor $\dfrac{X}{G}$ can be obtained as

$$\frac{X}{G} = \frac{\left[\left[(1-\beta^2) + 4\,\beta^2\xi\,\bar{\xi}\right]^2 + \left[2\bar{\xi}\beta\,(1-\beta^2) - 2\xi\beta\right]^2\right]^{\frac{1}{2}}}{(1-\beta^2)^2 + (2\bar{\xi}\beta)^2}$$

where $\beta = \dfrac{\bar{\omega}}{\omega}$, $\xi = \dfrac{c_{pass}}{2m\omega}$ and $\bar{\xi} = \dfrac{c_{pass} + c_{act}}{2m\omega} = \xi_{pass} + \xi_{act}$.

---

* Graduate Student, Dept. of Civil Engineering, University of California at Berkeley, CA.
** Professor, Dept. of Civil Engineering, University of California at Berkeley, CA.

Figures 1 and 2 compare the response of the system under passive damping and active damping. As expected the difference increases as β increases and as the value of ξ_{pass} and ξ_{act} increase. This tells us that for active isolation it is very convenient to design the passive isolation system as *flexible* as possible ($2<\beta<5$, $3secs < T_{fund} < 6secs$) in order to be far from all ground motion frequencies and further more to be able to increase the value of ξ_{act} as much as possible maintaining c_{act} which may be constrained to be smaller than a design value. Clearly an active scheme could allow us to introduce a force proportional to the absolute position of the mass and additional advantages could be obtained in the dynamic performance of the structure.

As we can notice as the system is more flexible or the active damping is increased the relative displacement at the isolator level tends to be equal to the ground displacement. The maximum displacement of the ground then becomes a very important parameter in the design process of an active isolation system since the isolators have a maximum allowable deformation.

State estimation is a crucial ingredient in this approach. Although absolute velocities can be satisfactorily estimated by integration of accelerations, a second integration will not give in general a sufficiently accurate estimate of the absolute position. This problem could be overpassed by means of the use of velocity seismometers which measure absolute velocity.

Clearly the higher the available control force the better the control goal can be obtained. However the maximum force is in general a constrained design variable. Reasonable control forces should not exceed in principle 2% of the weight of the building. This constraint is closely related to the controllability of the actuator and to the maximum total volume available in accumulators. The controllability of the actuator is dominated by the capacity of the servovalve and specifies for which range of relative velocities of the piston the force is satisfactorily controlled. Active isolation will in general assure very small absolute displacements of the superstructure and consequently the relative displacement if the piston of the actuator will be governed by the motion of the ground if the reaction of the actuator is applied on the ground or by the motion of the auxiliary mass if the reaction of the actuator is applied on an AMD. In both cases the volume of oil consumed during the shaking is linearly proportional to the area of the piston which given a pressure source (accumulators) defines the maximum force that can be delivered.

If the actuator is not fast enough time delay may be important and stability of the control strategy may be lost. If the actuator is grounded maximum ground velocities become an important design variable. If the actuator is attached to an AMD the mass of this system will determine the maximum velocities at which the AMD system will operate without problems in the control of the command force.

2. Experimental approach

The purpose of this section is to introduce the testing on two active isolation techniques which has been developed at the Earthquake Engineering Research Center of the University of California at Berkeley. The overall purpose of this research program is to evaluate the performance of several theoretical control strategies by their implementation in a reduced scale model.

3.1 Description of apparatus

A 4-story base isolated test frame (5 kips) was used for the experiments. Modeled as a 4 DOF system the frequencies of the system were identified as .6 Hz, 3.9 Hz, 11.9 Hz and 24 Hz. Figure 3 shows the test frame. The control force was delivered by a single electro-hydraulic actuator commanded by a 2.5 GPM servovalve. A digital controller was used to define the command control force and a digital proportional plus derivative scheme was used to control the force in the actuator. The software used in the experiments was developed by Automatic Testing Systems (ATS) and has the capability of controlling up to four actuators and recording data from sixteen channels. A 12 bit board was used for the control system. This configuration constrained the digitalization of the analog signals to the range of -2047 to +2048 bytes.

In this initial stage of the program the absolute velocities and displacements were measured directly using displacement and velocity transducers attached to a external fixed reference frame.

The actuator used in the experiments connected to a source pressure line of 3000 psi could exert a maximum force of 2 kips. The maximum forces required by the implemented controllers were significantly smaller than the capacity of the actuator. In this respect the design of the system was not optimized since a smaller actuator connected to the same pressure source and commanded by the same servovalve would have had a faster response and consequently the command control force would have been applied with smaller delays and higher accuracy.

3.2 Active damping

This active scheme is intended to reduce accelerations and absolute displacements in a base isolated structure by applying a force on the first floor , u_1, defined by the absolute velocity of the first floor, $\dot{x}_1$. This non linear control is defined as

$$u_1 = -\alpha \dot{x}_1 \ , \ \text{if} \ \dot{x}_1 < v_m$$

$$u_1 = -\alpha v_m \frac{\dot{x}_1}{|\dot{x}_1|} \ , \ \text{if} \ \dot{x}_1 \geq v_m$$

The maximum control force, $|u_{max}| = \alpha v_m$, did not exceed 212 lb. α was defined as 201 (lb/in/s).

3.3 Robust control

A control strategy proposed by A. Soldatos and G. Leitmann for base isolated structures called robust control was implemented on the test frame. The control force, $u_1(t)$, is applied on the first floor of a base isolated structure and assures that provided that the input ground motion satisfies $g(t) \leq y_{max}$ and $\dot{g}(t) \leq \dot{y}_{max}$ and the system matrix A is stable (usual situation), then the state $x(t)$ (absolute coordinates) is uniformly bounded with respect to some set S. The control is defined as

$$u_1 = (\frac{-B^T P x}{|B^T P x|}) \rho \ , \ \text{if} \ |B^T P x| \geq \varepsilon$$

$$u_1 = (\frac{-B^T P x}{\varepsilon}) \rho , \ \text{if} \ |B^T P x| < \varepsilon$$

where $\mathbf{B}^T = [\, 1/m_1 \; 0 \; 0 \cdots \; 0 \,]$, $\mathbf{P}$ is the symmetric positive definite solution of the Liapunov equation $\mathbf{A}^T \mathbf{P} + \mathbf{P} \mathbf{A} + \mathbf{Q} = 0$ for any definite positive definite $\mathbf{Q}$, and $\rho = \left[(k_0 g_{max}^2) + (c_0 g_{max}^2) \right]^{\frac{1}{2}}$. k_0 and c_0 are the stiffness and damping constants of the isolator. The set S can be made arbitrarily small by a small enough choice of $\varepsilon > 0$.

3.4 Results

Both active strategies were implemented and tested under simulated ground motions. The response of the passively base isolated structure was obtained under the same simulated ground motions for comparison. Both active isolation schemes gave significant improvements in the performance of the model. Figures 5a and 5b show the improvement obtained in the performance of the model when active isolation is combined with passive isolation. Accelerations and absolute displacements were drastically reduced compared with those of the passively base isolated model. Interstory drifts in the superstructure were reduced too. The active damping scheme did not introduce high frequencies into the model. However, high frequencies were introduced in the superstructure by the robust control scheme. This might not be convenient for sensitive secondary systems attached to the main structural system. This problem can be explained by the fact that the control force was applied with significant time delay.

REFERENCES

[1] Kelly J., Leitmann G. and Soldatos A., "Robust Control of Base-Isolated Structures under Earthquake Excitation", Journal of Optimization Theory and Applications, Vol.53, No 2, May 1987.

[2] Fujita T., Feng Q., Takenaka E.,Takano T.,Suizu Y., "Active Isolation of Sensitive Equipment for Weak Earthquakes", Proceedings of Ninth World Conference on Eartquake Engineering, Tokyo, August 1988.

[3] Soong T.T., "Active Structural Control: Theory and Practice", National Center for Earthquake Engineering Research, Buffalo, 1989.

[4] Inaudi J., Kelly J., "Active Control on Base-Isolated Structures", CE 299 Report, Department of Civil Engineering, University of California, Berkeley, May 1990.

[5] Rodellar J., Barbat A., Martin-Sanchez J., "Predictive control of Structures", Journal of Engineering Mechanics, Vol. 113, No 6, 1987.

[6] Aiken I., "Earthquake Simulator Testing and Analytical Studies of Two Energy Dissipation Systems for Multistory Structures", PH.D. Dissertation, University of California, Berkeley, October 1990.

[7] Richter, P. J., Nims, D. K., Kelly, J. M., and Kallenbach, R. M., "The EDR-Energy Dissipating Restraint, A New Device for Mitigating Seismic Effects ", Proceedings of the 1990 Structural Engineers Association of California Convention at Lake Tahoe, September 1990.

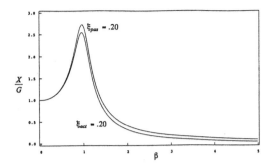

Figure 1. Amplification factor. Active damping vs. passive damping.

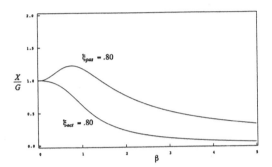

Figure 2. Amplification factor. Active damping vs. passive damping.

Figure 3. Test model. Isolation with linear springs.

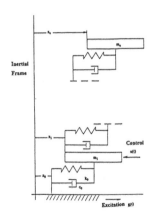

Figure 4. Active isolation.

p 129

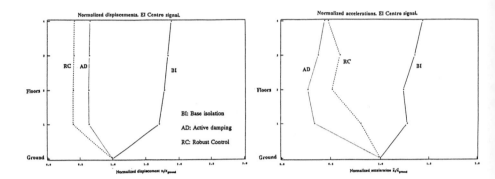

Figure 5a. Active isolation vs. passive isolation under simulated ground motion.

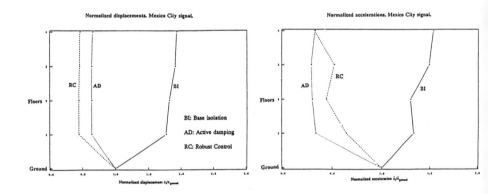

Figure 5b. Active isolation vs. passive isolation under simulated ground motion.

p 130

Some Issues Related to Active Control Algorithms

Wilfred D. Iwan and Zhikun Hou

Division of Engineering and Applied Sciences

California Institute of Technology

Introduction

Active control has recently become a focus of civil structure research due to its potential for providing a cost-effective means of improving the safety and functionality of certain structural systems. A significant effort has been directed towards the development of devices, algorithms and practical applications of active control. For a detailed review, the reader is referred to Soong (1988) and Kobori (1988).

While most previous studies of active control have been based on classical control theory as applied to highly idealized systems, this paper presents some results of a research effort which have concentrated on three fundamental and practically important aspects of the problem, i.e. acceleration control, optimum location of control devices, and the effect of time delays. In the first two areas, stochastic control theory has been employed to render the results useful not only for a particular sample of an environmental load, say a particular past earthquake, but also for events which may be modelled by a stochastic process.

Acceleration Control

In classical close-loop feedback control, the feedback control force is normally assumed to be a sum of terms proportional to the response displacement and velocity. However, in most civil structures, the response variable which can be most directly measured is acceleration. Therefore, the measured acceleration data must be integrated in the classical approach to obtain the necessary control force. This increases on-line calculation time and introduces additional system noise. The possibility of employing a control force which is directly propotional to measured acceleration has been explored for a simple control system.

Assume the differential equation of motion of the system be

$$m\ddot{x}(t) + c\dot{x}(t) + kx(t) + F(t) = f(t) \tag{1}$$

where m, c, k are mass, damping, and stiffness of the uncontrolled system respectively. The close-loop feedback control force $F(t)$ may be expressed in general form as

$$F(t) = \alpha\ddot{x}(t) + \beta\dot{x}(t) + \gamma x(t) \tag{2}$$

The special case of Eq. (2) where α is zero is the type of control force normally used. The case of nonzero α and zero β and γ is refered to as *acceleration control*. Similarily, velocity control, displacement control, and their combination may be defined. The load $f(t)$ is modelled as an modulated white noise. The control strategy is to choose appropriate α, β, and γ which minimizes a stochastic quadratic performance index defined as

$$J = \frac{1}{2} \int_0^{t_f} R_e E[\omega_0^2 x^2 + \dot{x}^2] + R_f E[F^2]dt \tag{3}$$

where t_f is the duration considered and R_e, R_f are weighting coefficients.

Conclusions can be drawn from some representative results shown in Fig. 1 and Fig. 2. Figure 1 gives a comparison of time histories of the mean square displacement S_{xx} by using the acceleration control, velocity control, displacement control, and velocity-displacement control respectively. A Sato-Shinozuka envelope is used in this example. The dashed line represents the result for the case without control. It is observed that acceleration control can be used effectively as conventional velocity-displacement control to suppress the structral response , but no on-line integration is required to obtain the control force. Figure 2 shows the results for the maximum S_{xx} versus the mean square of the control force required. Acceleration control may require a greater control force but the defference is probably reasonable. The conclusion indicated here can be extented to more complicated systems subjected to more general excitations.

Optimum location of control devices

In practice, only a limited number of control devices can be used to supress the response of a structure. With a limited set of control devices, consideration must be given to the optimal location of these devices. Since one does not know in advance the nature of the excitation to which the structure will be subjected, a statistical approach is warranted in determining optimal device location. In addition, the complex configuration of many civil structures may not allow to determine these positions intuitively.

An algorithm is developed to determine the optimum locations of control devices for a given number of devices and a given structural configuration. The governing equation of the structure can be written as

$$\mathbf{M}\ddot{\mathbf{X}}(t) + \mathbf{C}\dot{\mathbf{x}}(t) + \mathbf{K}\mathbf{X}(t) + \mathbf{F}(t) = \mathbf{f}(t) \tag{4}$$

which is a vector version of Eq. (1). The control force $\mathbf{F}(t)$ can be expressed as

$$\mathbf{F}(t) = \mathbf{U}\mathbf{F}_c(t) \tag{5}$$

where each component of $\mathbf{F}_c(t)$ is the control force contributed from a corresponding individual control device, and $\mathbf{U}$ is a *location matrix* whose entries are 1's or 0's indicating the location of the control devices. For any given matrix $\mathbf{U}$, a curve of mean-square control force applied and performance index or mean-square displacement response can be obtaind by using the independent modal space control method (Meirovitch, 1983). The method has been modified by eliminating a restriction that the number of modes employed in the analysis has to be equal to the number of devices used. The modal performance index is evaluated by the simplified space-variable approach developed by the authors (Iwan and Hou 1989, Hou 1990). Finally, the optimum location is determined by comparing the optimal values of performance index for different placement of devices, or equivalently, for different assignment for entries of matrix $\mathbf{U}$.

The algorithm has been tested for a simple case of a five-story building with equally distributed stiffness and damping properties. One actuator is used for illustration. The results are presented in Fig. 3 where roman numbers indicate locations where the actuator is placed. It is observed that a comparatively smaller mean-square control force is required for the same amount of mean-square displacement of structural response by placing the device on the fifth floor, and therefore, this is the best location for the control device.

Effects of time delay

In a real active control system, time delay may result from several sources including on-line computation, and the finite response time of certain physical components. Some investigators have performed laboratory studies of the effects of time delay and compensation methods have been proposed (Soong, 1988). Some numerical techniques have also been employed based on the assumption of small time delay (Abdel-Rohman, 1985). It is felt that the fundamental aspects of the problem still need to be investigated, which is justified by some results from the author's initial research effort directed toward the study of the dynamic behavior of linear systems with time delays using analytical techniques.

Consider the steady-state response of a simple linear system subjected a delayed feed-back control force expressed by

$$F(t) = \alpha \ddot{x}(t - \tau) + \beta \dot{x}(t - \tau) + \gamma x(t - \tau) \tag{6}$$

where τ is the time delay and the coefficients α, β, γ give a measurement of level of the control force. A representive curve for peak amplitude response versus nondimensional time delay $\frac{\tau}{T}$ is given in Fig. 4 where T is the natural period of the uncontrolled system.

It is observed that there exist critical values of the time delay for which the structural response becomes unbounded. The critical values may occur even for small ratio of the delay time to natural period. A more detailed study reveals that the number of critical values are, in fact, infinite, if they exist. These critical values may be classified into independent sets in which the critical values occur periodically. The number of the independent sets may be 0,1,or 2, as determined by the level of control force and the system damping ratio ς. Fig. 4 provides an example for which the number of independent sets is 2.

Though these observations are made for the steady-state response of a simple system, it shows that time delay effects may not be negligible in a control algorithm even though the time delay is small comparing to the natural period of the system. The initial results suggest that further study of this area is warranted.

References

1. Abdel-Rohman, Mohamed, "Structural Control Considering Time Delay Effect," Transaction of the CSME, Vol. 9, No. 4, 1985, pp. 224-227.

2. Iwan, W.D. and Hou, Z.K., "Explicit Solutions for the Response of Simple Systems Subjected to Nonstationary Excitations," Structural Safety, 6(1989), pp. 77-86.

3. Hou, Z.K., "Nonstationary Response of Structures and Its Application to Earthquake Engineering," California Institute of Technology. EERL 90-01, 1990.

4. Kobori, T., "State-of-the-Art Report, Active Seismic Response Control," Proceedings of Ninth World Conference on Earthquake Engineering, August 2-9, 1988, Tokyo-Kyoto, Japan. (Vol. VIII)

5. Meirovitch, L. and Silverberg, L.M., "Control of Structures Subjected to Seismic Excitations," ASCE, Journal of Engineering Mechanics, 109, pp. 604-618, 1983.

6. Soong, T.T., "State-of-the-Art Review: Acitve Control in Civil Engineering." Engineering Structures, Vol. 10, pp. 74-84, 1988.

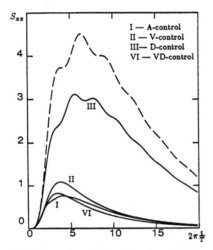

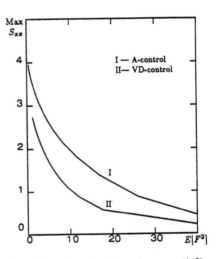

Fig.1: Comparison of S_{xx} using A-control, V-control, D-control, and VD-control. $\varsigma=0.1$. Dashed line represents the result for the case without control.

Fig.2: Comparison of maximum S_{xx} versus $E[F^2]$ using A-control and VD-control. $\varsigma=0.1$.

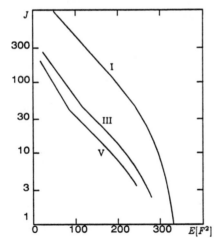

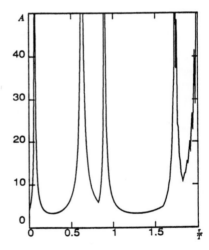

Fig.3: Comparison of results for objective func-tion J versus $E[F^2]$ by placing the actua-tor on 1st, 3rd, and 5th floor respectively Roman numbers indicate the position of the actuator.

Fig.4: Peak steady-state amplitude response A versus nondimensional time delay param-eter $\frac{\tau}{T}$. T is the natural period of the system.

Uniqueness in Structural System Identification

Lambros S. Katafygiotis and James L. Beck
California Institute of Technology, Pasadena, CA 91125

Abstract

A methodology is presented to investigate global identifiability within a class of structural models to provide the set of models within the given class which are equivalent as far as their input-output dynamic behavior is concerned. As an application, all the equivalent most probable models for given input and output data can be found, and they can then be used to give improved response predictions along with the associated uncertainty.

Introduction

The objective of system identification in structural dynamics is to improve mathematical models of the dynamics of a structure by using its measured response and possibly its measured excitation. The improved models are most often used for response prediction for possible future loads on the structure, for developing control strategies, or for damage detection and assessment.

Consider an appropriate class of models describing the input-output behavior of a structure, whose input and output have been measured over some time interval. One goal of our research is to use these data to predict the output of the system to other input. It has been shown that predictions can be made using the most probable model based on the data, assuming it is globally identifiable [Beck 1990]. This choice is not only the most rational one for choosing a single model from the class, it is also asymptotically correct for a sufficiently large number of sample datapoints, which is usually the case with dynamic tests or earthquake records of structural response. If the most probable model is not identifiable, predictions can be made using all those models in the specified class which are equivalent in the sense that they all give the same maximum probability based on the data [Beck 1990]. For a sufficiently large number of sample datapoints, all equivalent most probable models based on these data lead to the same predictions at the observed degrees of freedom, but this is not true for predictions at the unobserved degrees of freedom, and therefore, exploring this identification issue is particularly important when dealing with predictions at these degrees of freedom. In addition, it is easy to see the importance of exploring the nonuniqueness of the most probable model in system identification if the objective is control or damage detection and assessment.

Assuming that a most probable model has been identified by maximizing the probability based on the data, a methodology has been developed to find all other models, if any, in the specified class of models which give the same response at the observed degrees of freedom as the identified most probable model. This is accomplished by methodically searching the parameter space to find all models with the same modal frequencies and the same modal effective participation factors at the observed degrees of freedom. In the following section, a specific application is addressed to help clarify ideas.

Shear Structure Model

Assume a linear N-story planar shear structure with classical modes, with base excitation and starting from rest, as in Figure 1. The equation of motion is

$$M\underline{\ddot{x}} + C\underline{\dot{x}} + K\underline{x} = -M\underline{b}\ddot{z}(t); \quad \underline{\dot{x}}(0) = \underline{0}, \underline{x}(0) = \underline{0} \tag{1}$$

where $\underline{b} = [1, 1, \ldots, 1]^T$. The response at the ith floor can be expresed by using modal analysis as follows

$$x_i(t) = \sum_{r=1}^{N} p_i^r \xi_r(t) \tag{2}$$

where

$$\ddot{\xi}_r + 2\zeta_r \omega_r \dot{\xi}_r + \omega_r^2 \xi_r = -\ddot{z}(t) \quad ; \quad \xi_r(0) = 0, \dot{\xi}_r(0) = 0 \tag{3}$$

and

$\underline{\phi}^r : r^{th}$ modeshape normalized so that $(\underline{\phi}^r)^T M \underline{\phi}^r = 1$

$\omega_r : r^{th}$ modal frequency

$\zeta_r : r^{th}$ modal damping factor

$\alpha_r : r^{th}$ modal participation factor

$p_i^r = \alpha_r \phi_i^r :$ effective participation factor of the rth mode at the ith degrees of freedom.

It has been shown [Beck 1978], that there is a one to one mapping between the set of possible model responses $x_i(t)$ for a given finite-duration input, and the set of possible modal parameters $\{\omega_r, \zeta_r, p_i^r : r = 1, \ldots, N\}$, that is, these modal parameters are globally identifiable.

Assume that the floor masses, and therefore the mass matrix M, are known, while the damping matrix C can be constructed from the damping ratios ζ_i, which as previously mentioned, can be identified uniquely. The stiffness matrix is assumed unknown and parameterized by $\underline{\theta} \in \Theta$. The interstory stiffness between the $(i-1)^{th}$ and i^{th} floor is assumed to be $K_i = \theta_i k_i$, where the k_i's are prescribed while the normalized parameters $\theta_i, i = 1, \ldots, N$ are to be identified. The class Θ is identifiable if and only if there is a one-to-one mapping between $\underline{\theta}$ and $\Omega = \{\omega_r, p_i^r : r = 1, \ldots, N, i \in S_o\}$ where S_o is the set of observed degrees of freedom.

In the following section , an algorithm is presented to check identifiability of $\underline{\theta}$ by methodically searching the Θ space to find all other $\underline{\theta}$, if any, corresponding to the same set of modal parameters Ω.

Algorithm

Given $\underline{\hat{\theta}}^1$ corresponding to $\hat{\Omega} = \{\hat{\omega}_r, \hat{p}_i^r : r = 1, \ldots, N, i \in S_o\}$, the Θ space is explored to find the whole set $\Theta_{\hat{\Omega}} = \{\underline{\hat{\theta}}^k, k = 1, \ldots, N_{\hat{\Omega}}\} \subset \Theta$, of $\underline{\theta}$'s corresponding to the same set of modal parameters $\hat{\Omega}$. If $N_{\hat{\Omega}} = 1$ then $\underline{\hat{\theta}}^1$ is said to be globally identifiable. It has been shown [Udwadia et al 1978], that for $S_o = \{N\}$, that is, if only the response of the roof is measured,then $N_{\hat{\Omega}} \leq N!$. If $\Theta_{\hat{W}} = \{\underline{\hat{\theta}}^l, l = 1, \ldots, N_{\hat{W}}\} \in \Theta$ is the set of $\underline{\theta}$'s corresponding to the same set of modal frequencies $\hat{W} = \{\hat{\omega}_r, r = 1, \ldots, N\}$, it can also be shown that $N_{\hat{W}} \leq N!$. Since $\Theta_{\hat{\Omega}} \subseteq \Theta_{\hat{W}}$ and $\Theta_{\hat{W}}$ is of finite dimension, it is easy to obtain $\Theta_{\hat{\Omega}}$ from $\Theta_{\hat{W}}$ by just discarding the $\underline{\theta} \in \Theta_{\hat{W}}$ which do not correspond to $\{\hat{p}_i^r : r = 1, \ldots, N, i \in S_o\}$.

If $\Theta_{\hat{\omega}_i} = \{\underline{\theta} \in \Theta : \omega_i(\underline{\theta}) = \hat{\omega}_i\}$ is the $N - 1$ dimensional hypersurface of constant i^{th} modal frequency, then $\Theta_{\hat{W}} = \cap_{i=1}^{N} \Theta_{\hat{\omega}_i}$. It can be seen that $\tilde{\Theta}_{\omega_k} = \cap_{\substack{i=1 \\ i \neq k}}^{N} \Theta_{\hat{\omega}_i}$, the set of $\underline{\theta}$ corresponding to $\omega_i(\underline{\theta}) = \hat{\omega}_i, i = 1, \ldots, N, i \neq k$, is a one-dimensional curve within the N-dimensional space Θ. By following all the different curves $\tilde{\Theta}_{\omega_k}$ passing through $\underline{\hat{\theta}}^1$, and monitoring when, if ever, the "released" frequency ω_k corresponding to $\underline{\theta} \in \tilde{\Theta}_{\omega_k}$ becomes equal to $\hat{\omega}_k$, the equivalent solution set $\Theta_{\hat{W}}$ is built up. The algorithm has a systematic way of following all possible curves $\tilde{\Theta}_{\omega_k}$ passing through the different equivalent solutions. The concepts of the algorithm are illustrated in Figure 2 for the two-degree of freedom case.

Application

The algorithm presented previously was applied to find the set $\Theta_{\hat{W}}$ of a uniform N-story building , that is, $\hat{\underline{\theta}}^1 = [1, 1, \ldots, 1]^T$ is one solution if $k_1 = k_2 = \ldots = k_N$. It was found that if $S_0 = \{N\}$, that is, if only the roof response is observed, then $\Theta_{\hat{W}} = \Theta_{\hat{\Omega}}$,so that all $\underline{\theta}$ giving the same N modal frequencies also have the same modal effective participation factors at the roof. Also, for the shear buildings equivalent to the uniform one, it was found empirically, that $N_{\hat{W}} = 2^M$ where M is the integral part of $N/2$, so for large N, $N_{\hat{W}}$ is much smaller than the previous upper bound $N!$.

Table 1 displays the equivalent parameter sets $\underline{\theta}$ for a 6-story uniform shear structure. Figure 3 shows the effective participation factors of the different modes at the different story levels for these equivalent models and Table 2 displays the statistics of their variation, assuming the models are equally probable a priori. It is interesting that all equivalent models have the same effective participation factors at the roof while at lower floors the discrepancies increase. Therefore, it is obvious that by using a single optimal model based on the base acceleration and roof response ,the predictions will be unreliable at the lower floors. The actual uncertainty in the predicted response $x_i(t)$ can be computed using (2) and Table 2, since $x_i(t)$ is a linear combination of the uncertain, but equally probable, effective participation factors. Note that the modal coordinates $\xi_r(t)$ in (2) are the same for all equivalent models.

Conclusion

An algorithm to investigate global identifiability for model identification in structural dynamics is presented. It is shown that choosing just a single model, as usually done by estimating the model parameters through optimization of the model and measured responses at certain degrees of freedom, can lead to unreliable response predictions at the unobserved degrees of freedom, when the model used is not globally identifiable.

References

1. Beck, J.L., "Determining Models of Structures from Earthquake Records," EERL Report No. 78-01, California Institute of Technology, 1978.

2. Beck, J.L., "Statistical System Identification of Structures," Structural Safety and Reliability, ASCE, II, 1395-1402, 1990.

3. Udwadia, F.E., Sharma, D.K., Shah, P.C., "Uniqueness of Damping and Stiffness Distributions in the Identification of Soil and Structural Systems," Journal of Applied Mechanics, 45, 181-187, 1978.

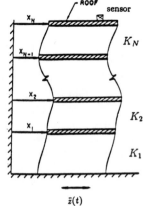

FIGURE 1 Shear Building Model

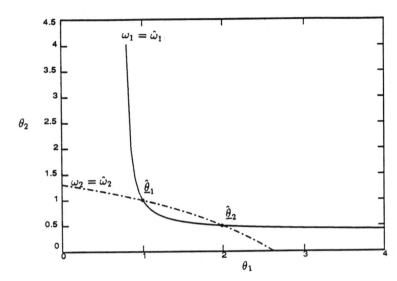

FIGURE 2 Constant Frequency Curves in the Parameter Space for a 2-Degree-of-Freedom Uniform Shear Building

No.	θ_1	θ_2	θ_3	θ_4	θ_5	θ_6
1	1.0000	1.0000	1.0000	1.0000	1.0000	1.0000
2	1.5848	0.6963	1.2875	0.7574	1.1766	0.7898
3	1.9970	0.7980	0.7095	1.3848	0.7113	0.8980
4	2.0000	1.0000	1.0000	0.5000	1.0000	1.0000
5	2.0932	1.0476	0.7240	0.7374	0.6705	1.2738
6	2.2911	0.6304	0.9321	1.1774	0.9515	0.6631
7	2.4913	0.8777	0.6514	1.1106	0.6672	0.9475
8	2.8252	0.6753	0.8826	0.9021	0.8753	0.7520

TABLE 1 Equivalent Sets of θ's for 6-story Uniform Shear Building

Floor	Mode 1	Mode 2	Mode 3	Mode 4	Mode 5	Mode 6
1	0.154±40%	0.146±43%	0.139±44%	0.156±91%	0.166±122%	0.240±95%
2	0.491±16%	0.380±17%	0.253±46%	0.076±93%	-0.061±143%	-0.139±70%
3	0.776±8%	0.391±22%	-0.038±144%	-0.175±47%	-0.045±135%	0.090±35%
4	1.008±3%	0.161±37%	-0.242±22%	0.021±305%	0.111±54%	-0.059±38%
5	1.175±1%	-0.163±25%	-0.085±59%	0.139±31%	-0.096±26%	0.030±24%
6	1.258±0%	-0.379±0%	0.183±0%	-0.090±0%	0.038±0%	-0.009±0%

TABLE 2 Mean and Coefficient of Variation of Effective Participation Factors

p 139

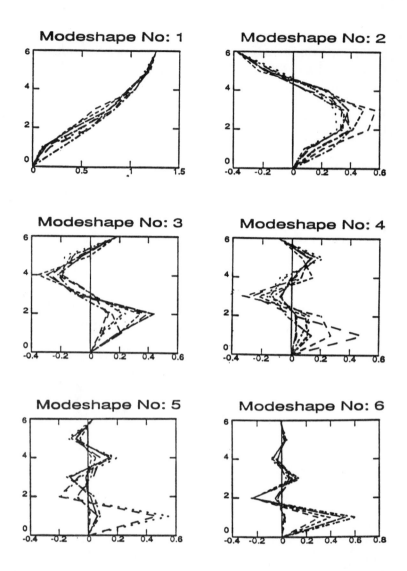

FIGURE 3 Effective Participation Factors corresponding to the Equivalent Sets of θ's for 6-story Uniform Shear Building

STRUCTURAL RESPONSE CONTROL TECHNOLOGIES OF TAISEI CORPORATION

by

Hiroshi Kitamura[*1], Soichi Kawamura[*2],
Masaaki Yamada[*3] and Shunji Fujii[*4]

INTRODUCTION

TAISEI has already established a sliding-type base-isolation system named "TASS (TAISEI'S Shake Suppression)" system. Three buildings were actually built using TASS system, and the effect of response acceleration reduction is confirmed by the earthquake motion measurement of those buildings. A floor isolation system named "TASS FLOOR" has also been developed using sliding mechanism, and installed in some buildings with computer rooms.

This paper describes two types of response control technology which have been developed by TAISEI other than TASS system. The one is the variabe friction bearing (VFB) system which uses hydropressure to minimize the coefficient of friction of slide-bearing. The other one is gyroscopic stabilizer which makes use of precession motion of gyro to suppress the bending deflection of tower-like structures. Both systems were thought out as passive-type systems initially, but extension to active or hybrid systems is intended. Patents are claimed for both systems.

---

*1   Senior Executive Director, Manager of Technology Division, Eng. Dr. TAISEI CORPORATION, JAPAN,

*2   Leader of Wind & Earthquake Engineering Group, Technology Research Center, Eng. Dr.

*3   Chief Researcher of Wind Engineering Team, Technology Research Center, Eng. Dr.

*4   Chief Engineer, Technology Development Department, Eng. Dr.

[1] VARIABLE FRICTION BEARING (VFB)

1. COMPOSITION AND MECHANISM

Variable friction bearing (VFB) is a kind of friction bearing which makes use of hydropressure to control the coefficient of friction. The composition and mechanism of the Variable Friction Bearing (VFB) is schematically shown in fig. 1 compared with oridinary sliding bearing. The VFB is a carved-out steel disc with a slide-ring on its surface and a fluid chamber in the interior. The inside of the slide-ring is sealed and the chamber is filled with pressurized liquid such as oil or water. Out of the total normal force N, a part N_1 is carried by the slide-ring and the other part N_2 is carried by the liquid inside the chamber. While the frictional force due to the liquid is virtually zero, frictional force due to the sliding bearing is $\mu \times N_1$ or $\mu \times (N-N_2)$, which is substantially smaller than the frictional force $\mu \times N$ of an ordinary sliding bearing, where μ is the coefficient of friction.

By using VFB for a base isolation system, the isolation effect can be enhanced remarkably compared with the ordinary isolation method and the seismic force input to the building can be substantially reduced. Other advantage of VFB is that the control as well as the reduction of the frictional force is possible. It can also work as braking system when pressure is lowered or negative pressure is supplied. This is effective in the application to the active or hybrid response control system.

2. BASIC EXPERIMENT

To verify the dynamic behaviour of VFB its model was set up and subjected to lateral sinusoidal load under constant normal load. The outer diameter of the bearing used in the test is 23.5cm at the contact surface. Interior pressurized zone is 19.5cm in diameter. The applied pressure varied from 0 to 80 kg/cm². Pressure control system used in the test is shown in fig.2. Pressure control was triggered by accelerometer and displacement gauge. When the acceleration of the base exceeds 30 Gal, pressure is automatically applied, and if the slide displacement exceeds 3 cm, the pressure is reduced to zero by pressure release valve.

Based on the experiment the performance of VFB can be summarized as follows :

1) As fluid pressure goes up, coefficient of friction goes down almost linearly toward the limit value which corresponds to the resistance of the sealant as shown in Fig.3. Hysteresis loops gets thinner as applied pressure goes up as shown in Fig.4.

2) O-ring is effective in sealing pressurized fluid up to 80 kg/cm^2 as long as the slide-ring is touching the bearing plate.

3) Programmed control works good and the reaction time is short. Hence the VFB can be used for the active response control system.

4) Automatic fluid feeding is efficient and pressure is maintained constant even when there is some leak of liquid.

3. CONCEPT OF RESPONSE CONTROL SYSTEMS USING VFB

Three types of response control systems are intended using VFB systems; active mass damper, hybrid base-isolation system and input cancelling system.

Fig. 5 (a) is an active mass damper with an auxiliary mass floated on sliding base by VFB.

Fig. 5 (b) shows the concept of hybrid base isolation system using VFB. When an earthquake ground motion is detected by sensors, VFBs are pressurized within fraction of a second. The coefficient of friction can be reduced to 0.02 or less, as was verified by the experiment. Hence, the seimic shear force of the building becomes less than 2 % of the weight of the building. In case of excessive sliding displacement, sliding of VFB can be limited by reducing pressure or by applying suction pressure. The reduction of sliding will result in increased input force to the building and this should be considered in the design stage.

Fig.5 (c) shows input cancelling system which is an actuator-controlled hybrid base isolation system using VFB. Seismic force acting on the building is not only mitigated by low coefficient of friction of VFBs but also compensated by actuators. Hence, the force

input to the building may be reduced as small as possible.
During small to medium earthquakes, actuators can be activated with
stable power supply and force input to the structure may be reduced to
nearly zero.   This is important from human comfort point of view.
During strong earthquakes, even when external power supply is
interrupted, severe structural damage can be avoided by VFBs
activated by reserved pressure in accumulators.

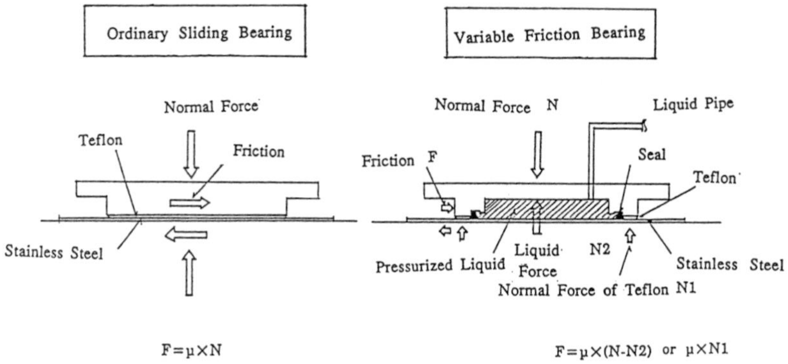

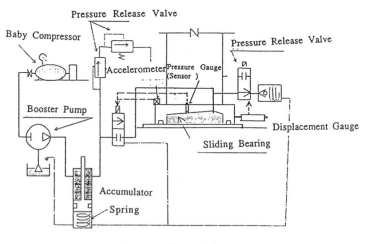

Fig.1 Mechanism of VFB

Fig.2 Pressure Control System

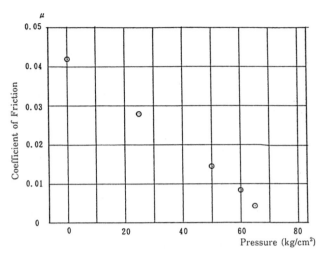

Fig.3 Pressure vs. Coefficient of Friction (N = 20t)

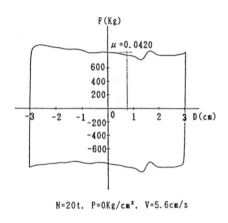

N=20t. P=0Kg/cm², V=5.6cm/s

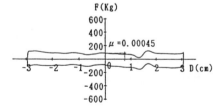

N=20t. P=65Kg/cm², V=5.6cm/s

Fig.4 Load-Displacement Curve (N = 20t, p = 0 and 65 Kg/cm²)

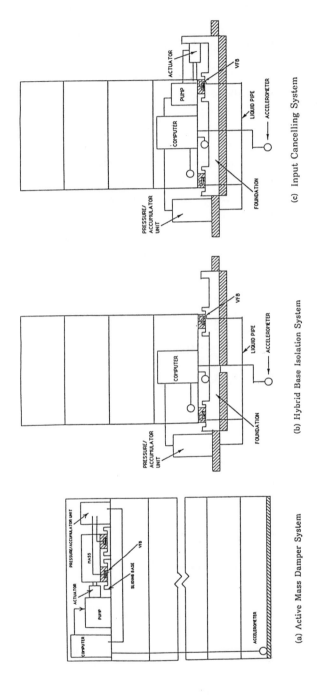

(a) Active Mass Damper System

(b) Hybrid Base Isolation System

(c) Input Cancelling System

Fig.5 Response Control Systems Using VFB

p 146

(2) GYROSCOPIC STABILIZER

1. COMPOSITION AND MECHANISM

Gyroscopic stabilizing method has been studied to limit the wind-induced vibration of tower-like structures within an acceptable range.

This system uses a high-speed rotating disk to provide a resistant moment to the bending or rocking motion of the structure.

Composition of the device is shown in Fig.1. High-speed rotating disk is encased in a holder and turns around Z-axis. The holder is fixed to the structure at two ends (A, B) and it can rotate freely around X-axis.

The bending deformation of a structure gives rotatinal motion Ω_Y to the stabilizer which causes the disk to rotate around X-axis. When this rotation Ω_X takes place, a large stabilizing moment M_Y, proportional to the angular velocity of the rotation as expressed by the following equation (1), is exerted on the structure.

$$M_Y = I_Z \cdot \omega \cdot \Omega_X \quad \dots\dots\dots\dots\dots\dots\dots \quad (1)$$

where I_Z : rotational inertia of spinning disk
   ω : angular velocity of the rotation of the disk around Z-axis
   Ω_X : angular velocity of precession (the rotation of the disk around X-axis)

This system requires the power to keep the disk spinning.

Fig.2 shows the deformation of a cantilever applied with external force P, stabilizing moment M_Y. Deflection δ of a cantilever with height h and bending rigidity EI can be expressed by equation (2). As shown in this equation, the stabilizing moment applied at the top reduces the horizontal deflection of the bending structure.

$$\delta = P \cdot h^3 / (3EI) - M_Y \cdot h^2 (2EI) \quad \dots\dots\dots \quad (2)$$

2. FUNDAMENTAL TEST AND ITS SIMULATION

Free vibration and wind-induced forced vibration tests were conducted using the test model as shown in Fig.3. Natural period of the model is

2.4sec, and the weight of the mass point is 420kg. The weight of gyro and its maximum speed of rotation is 3kg and 12,000 rpm respectively.

Fig.4 shows the results of free vibration test with initial deflection 10mm. Fig.5 shows the results of wind-induced forced vibration test using electric fan. As shown in these figures, horizontal deflection of the model could be greatly reduced by the gyro action compared with the case without gyro action. Orthogonal vibration was not induced.

Simulation analysis was conducted using the analytical model shown in Fig.6. Fig.7 shows the result compared with that of the free vibration test. Solid line shows the analytical result and dotted line shows the test result. As can be seen in these figures, analytical result agrees well with the experimental value.

3. EXTENSION TO HYBRID SYSTEM

Generally, an active control system is a feedback or feed-forward control system which is designed as to sense the structural or input motion and to generate a corrective control force acting on the structure in order to reduce the dynamic response. On the other hand, gyroscopic stabilizer is essentially a passive device and has no sensor and no control system by itself. It gives stabilizing moment due to the precession induced by the motion of the structure. Making the most of this effect, gyroscopic stabilizer can also be applied to hybrid system.

Angular velocity of precession Ω_x, which generates the stabilizing moment, is governed by the movement of the structure. And it is not so difficult to make the velocity faster as the movement of the structure is slow in general.

To increase the angular velocity Ω_x, active movement $\Delta\Omega_x$ is added by external motor. The external motor is controlled by pilot gyro which is very small and sensitive. Fig.8 shows the composition of the system. Soon after the movement of structure starts, the rotor of the pilot gyro responds, and this supplies current to the external motor. The motor starts to drive the stabilizing gyro. Thus, control of this hybrid system is very easy compared with those of general active control systems. And this system can supply larger stabilizing moment than that of the passive one.

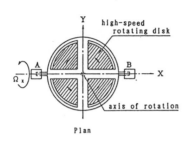

Plan

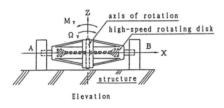

Elevation

Fig.1 Gyroscopic Stabilizer

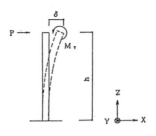

Fig.2 Force and Deflection of
a Cantilever

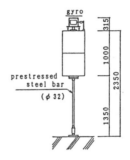

Fig.3 Test Model

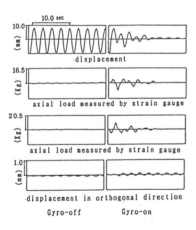

Gyro-off   Gyro-on

Fig.4 Results of Free Vibration Test
(initial deflection : 10mm)

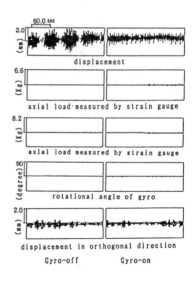

Gyro-off   Gyro-on

Fig.5 Results of Forced Vibration Test

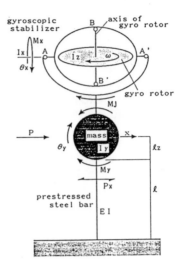

Fig.6 Simulation Model

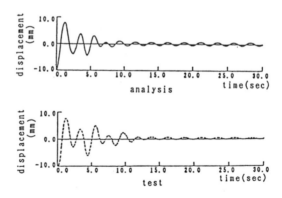

Fig.7 Analytical and Test Results of Free Vibration

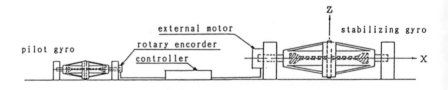

Fig.8 Composition of a Hybrid System

Seismic Response Controlled Structure with Active Mass Driver System and Active Variable Stiffness System

Takuji KOBORI[1], Mitsuo SAKAMOTO[2],
Motoichi TAKAHASHI[2], Norihide KOSHIKA[2], Koji ISHII[3]

1) Professor Emeritus of Kyoto University, Dr. of Eng.
Executive Vice President of Kajima Corporation
2) Kobori Research Complex, Kajima Corporation
3) Kajima Institute of Construction Technology

The Active Mass Driver (AMD) system, which is classified as a response control force type, aims to suppress building vibrations caused by earthquakes and strong winds by installing an auxiliary mass in the building and controlling it by operating an actuator. In succession, our studies were promoted for its practical use, and in August 1989, this AMD system was installed and put into operation in a building built in Tokyo. This system has been verified to achieve the purpose through observations during actual earthquakes and strong winds. The practical application of the active earthquake and wind-induced response controlled structure system is the very first of its kind in the world.

The Active Variable Stiffness (AVS) system is a nonresonant type seismic response control system to reduce the response of buildings by actively controlling the stiffness of structures in order to prevent the resonance with the uncertain earthquake motion that varies every moment.

These systems aims at ensuring the structural safety, improving the living comfort, and maintaining the function of various equipment including valuable information oriented appliances.

This paper reports, regarding the AMD system, on the outline of the composition of the AMD system, the experiments conducted to the building with vibrators upon its installation, and the earthquake records observed and obtained after its installation and use, based on reference [1]. And regarding the AVS system, it also describes the results of the experiment using a shaking table which was carried out with the purpose of applying it for actual system and for confirming its characteristics, based on reference [2].

1. ACTIVE MASS DRIVER SYSTEM

1.1 Outline of the Building

The AMD system was installed in an 11-story building, which was constructed in Tokyo. The typical floor plan and cross section are shown in Figure 1. This is a very slender building with dimensions of 4m front width, 12m length and 33m height. Regarding the structure of the building, above ground is with rigid framework of four structural steel box columns and I shaped beams. The total weight of the building estimated at the time of designing is about 400 tonf.

Initially in design process, an anxiety was felt towards vibrations during earthquakes and strong winds because the building has a very slender shape and the damping factor of the structural steel beam column framework is small. Thus, in order to reduce the vibrations and to satisfy the serviceability condition, it was decided to install the AMD system.

1.2 Purpose of Control

In Japan, we frequently experience the moderate earthquakes of a maximum acceleration of about 10 cm/s^2. The purpose of installment of this AMD system is to reduce the maximum response quantities of an uncontrolled structure to about one half through two third during the frequent earthquakes. In other words, it means improvement to the living comfort subjected to such the external excitations as earthquakes or strong winds.

1.3 Control Algorithm

The matrix equation of motion for the structure subjected to an external excitation such as an earthquake and wind can be expressed as

$$M\ddot{q}(t)+C\dot{q}(t)+Kq(t)= Uu(t)+f(t) \qquad (1)$$

in which q(t)= the displacement vector relative to the base, u(t)= the control force vector, f(t)= the external force vector, U= the location matrix of controllers, and M, C, and K= the mass, damping, and stiffness matrices, respectively.

In the control algorithm adopted in the system, the control force u(t) is regulated linearly by the relative velocity vector q(t), i.e.,

$$u(t)= G\ \dot{q}(t) \qquad (2)$$

where G is the feedback gain matrix that is, first, evaluated by the optimal control theory for all components of the state vector, that are relative velocities and displacements at each story. However, since it was at variance with reality to observe all responses at every story, the feedback gain were simplified and regulated by certain relative velocities so as to have an equivalent control effect to the effect obtained by the optimal control theory for all components. The control system was designed based on the simplified feedback gain so as to achieve the purpose of control mentioned in section 1.2.

1.4 Composition of the AMD System

The entire composition of AMD system for the building is shown in Figure 2. There are two units of weights that provide the controlling force, of which the center one (auxiliary weight of four tonf) acts to suppress the large lateral vibrations in the width direction, and the one on the end (auxiliary weight of one tonf) acts to suppress the torsional vibrations.

The vibrations of the building are detected by the sensors set inside the building. The signal then is transformed into operation commands by the control computer. In order to activate the reaction force which controls the vibration, the actuator operates the weight in accordance with the commands.

With the purpose of confirming the control effect and to obtain data for utilizing to development henceforth, an observation system is also installed in the building. The accelerations at the basement, 6th, and 11th floor are measured and also the accelerations and relative displacements of the weights are measured.

The AMD system is composed of the driver unit, hydraulic supply unit, control computer. The detail of the each components are described below.

Driver Unit
For realizing the objective of the control system, the required control force determined were one tonf for the lateral vibration and 0.25tonf for the torsional vibration. The driver unit

that was designed based on above conditions are shown in Figure 3. The driver unit consists of steel weight suspended by steel cables, the actuator, and the reaction block that connects the building and the actuator. The natural frequency of the suspended weights are approximately 0.4Hz which is much lower than building's natural frequency.

In order to install the equipment in a narrow space, the actuator is mounted above the weights, and operates the weights with the building as the reaction force. The actuator is an electro-hydraulic type of large power, and a servo mechanism is used. Control forces are applied accordingly with the feedback signal of acceleration and displacement of the weights.

Hydraulic Power Unit
The hydraulic power unit consists of 2 pumps, of 1.5kw and 22kw, the oil tank, filter, the accumulator and the fan cooler. The 1.5kw pump is placed in constant activation and provides pressure storage to the accumulator, and gives coverage to the rise of the 22kw pump which enables instant operation when an earthquake hits. Due to this dual pump system the electrical power conservation is possible.

Control Computer
The control computer system is made up of a control circuit of analog type, which creates the commands (by processing the signals from the sensors), and the controller for the actuator. The method of control applied is expressed by equation (2), and this equation is concretely as follows:

$$u_{AMD1}(t) = g_1 \dot{q}_1(t) + g_2 \dot{q}_{AMD1}(t) \qquad (3)$$

$$u_{AMD2}(t) = g_3(\dot{q}_2(t) - \dot{q}_1(t)) + g_4 \dot{q}_{AMD2}(t) \qquad (4)$$

where: $u_{AMD1}(t)$, $u_{AMD2}(t)$: control forces of AMD
$\dot{q}_1(t)$, $\dot{q}_2(t)$: relative velocities at the 11th floor
                  $\dot{q}1$; at the center, $\dot{q}2$; at the end
$\dot{q}_{AMD1}(t)$, $\dot{q}_{AMD2}(t)$: relative velocities of AMD weight
g_1, g_2, g_3, g_4 : feedback gain

In this system the vibrations of the building and weights are detected by acceleration sensors. These signals are transformed into the velocity signals by the control circuit, and the control commands are made with their signals. Therefore, the control forces are defined by equations of multiplying the transmitted velocities and feedback gains which was determined by the preliminary analysis.

1.5 Vibration Test

Test Method
Forced vibration test for the newly built building was conducted by using two electro-magnetic vibrators that have a capacity of 250 kgf exciting force at maximum and are placed at 10th floor shown in Figure 1. The two vibrators excited the building, in the state of AMD off and AMD on, in width direction with the same phase harmonic loadings and in torsional direction with the opposite phase harmonic loadings.

Dynamic Characteristics of Structure
As the test results, resonance periods and modal damping coefficients estimated from free vibration waveforms are indicated in Table 1.

A typical resonance curve of displacement measured at the 10th floor under the AMD off is shown in Figure 4. From the figures and table, apparent remarks are (1) vibration modes of width direction and torsional direction are excited independently, and (2) modal damping coefficients of torsional modes are larger than those of width direction modes.

AMD Performance Subjected to Stationary Excitation
A resonance curve of displacement measured at the 10th floor of the controlled structure under the AMD on is shown in Figure 5(a). This shows notable effect of control in comparison with Figure 4, especially at the resonance peak of 1.07Hz, first mode in width direction, and 1.85 Hz, first mode in torsional direction.

Figure 5(b) shows the control forces produced by AMD1 that performed remarkably to control the predominant modes in width direction and AMD2 that performed to control the predominant modes in torsional direction, as intended in the design.

1.6 Earthquake Observation Records

Five seismographs that can observe and record the accelerations in mainly the width direction during earthquakes were installed at the basement B1, 6th and 11th floor.

After the building was completed, all installed seismographs recorded the acceleration response of each floor of the building with the AMD on during several earthquakes. Among them, two earthquake records are picked up for verifying the effectiveness of AMD system. One is an earthquake of October 14, 1989 (Magnitude 5.9) and the other is of June 1, 1990 (Magnitude 6.0).

1.7 Simulation Analysis

Analytical Model of Building
The objective building is modeled as a 3-D frame that consists of columns, girders and joint panels. Precise deformations of each elements are considered and reduced to the horizontal displacements Ui, Vi, and the torsional displacement Φi at the gravity center that represent the i-th floor displacements from assumption of a rigid floor slab. The generalized stiffness matrix is derived from the transformation calculation of the local stiffness matrix and the compatibility conditions at each floor level. The frames and stiffness of partition walls around the elevator is also considered in addition to the main structural elements. The analytical model is fixed at the first floor level.

Natural periods and vibration modes obtained by the eigenvalue analysis of the analytical model are shown in Table 1 and Figure 6 in comparison with the test results. In the vibration modes, the mode shape was normalized relative to the amplitude of the 11th floor. Judging from these table and figures, the analytical model is appropriate to be used for the simulation analysis of vibration tests and earthquake response.

For AMD1 and AMD2, whose weight was modeled as a mass point respectively, the dynamic characteristics of control circuits and actuators in the frequency domain are considered in the analysis.

Simulation Analysis of Vibration Test
A resonance curve of displacement of the 10th floor, that was evaluated from the analysis using the analytical model that was excited by a force of 100kgf at the 10th floor, is shown in Figure 7. In this case, in order to correspond with the test condition, a horizontal exciting force of 100kgf operated in width direction near the resonance frequencies of horizontal vibration modes and an exciting couple force of 100kgf operated in torsional direction near the resonance frequencies of torsional vibration modes.

In comparison between the analytical results and test results, the analytical model that included the characteristics of control circuits and actuators is regarded as appropriate because the analysis simulated the experimental amplitude of the first width direction mode near 1Hz and first torsional mode near 1.8Hz that can be controlled mainly by AMD.

Simulation Analysis of Earthquake Responses

Based on the verified analytical model, in order to confirm the AMD performance during earthquakes, analyses were carried out to compare the responses of the uncontrolled and controlled structure. Figure 8 and 9 show the accelerations and the relative displacements of the 11th floor of the uncontrolled and controlled structure under the two earthquakes. A remarkable decrease of the amplitude due to the AMD system can be seen in each time history of the figure. Therefore, the AMD system performed as the authors prescribed during the earthquake.

1.8 Summary

In applying the Active Mass Driver system to the 11-story office building constructed in Tokyo, the feedback gain was determined based on the optimal control theory and simplified to be used in the control circuit. A vibration test under a stationary excitation verified the AMD effectiveness as prescribed in the design. And the simulation analysis that could simulate the earthquake responses recorded by the seismographs confirmed the AMD effectiveness under nonstationary excitations such as earthquakes.

From the results of the vibration test and the simulation analysis, the introduced system is verified to achieve the purpose of the vibration control.

2. ACTIVE VARIABLE STIFFNESS SYSTEM

2.1 Outline

The Active Variable Stiffness (AVS) system is classified as a nonresonant type seismic response control. The system actively controls the vibration characteristics of the building so that resonance with the continuously arriving earthquake motions can be avoided and the building's response can be suppressed. The model tests of this system have already been finished and it is now in the stage of practical application. In November 1990, this system will be realized to actual 3-story building.

2.2 Shaking Table Tests

Three Story Specimen
The test specimen is composed of three-story and one-span steel frames in excitation direction as shown in Figure 9. The variable stiffness device, namely cylinder lock device, was installed between the beam of each floor and the top end of the inverted V-shaped braces. The device is a two-ended rod-type hydraulic closing cylinder. A switch valve is installed in the connecting tube that joins two separate cylinder chambers. The joint between the brace and the frame is locked or unlocked by opening or closing the valve. The stiffness of the entire structure can be altered. The necessary energy for this operation is only the 20W of electricity to operate this switch.

Method of Control
In this system, in order to select the nonresonant vibration system, an analysis of the input ground motion was performed in real time by using a frequency analyzer. In the test, vibration systems to be selected are three types as shown in Figure 10.

Although this system can be adapted to various control algorithms, for the tests the efficiency of the feedforward control was verified. The control was performed in the following order.
(1) The predominant period of the input ground motion to the test specimen was analyzed in real time by the seismic motion analyzer (band-pass filter).
(2) A computer selected a stiffness type with less possibility of resonance.
(3) The system was switched over to be the selected stiffness type by the cylinder lock device.

Test Results

As typical examples, shown in Figure 11 are the accelerations and displacements at the top of the test specimen when the Mexico Earthquake was input. In those figures, the process of changing the structural stiffness can be readily seen under control. The responses of each stiffness type without control were reduced significantly by means of ingenious stiffness switch over.

2.3 Summary

The control efficiency of the nonresonant type AVS system was confirmed by the model tests and it is now in the stage of practical application. An excellent features of this system is that the system can operate adequately with only enough electricity for emergency sources, thus contributing to power energy saving. Therefore, unlike the response control force type devices, this system can operate during large earthquakes. Now, R & D is being promoted so that adequate effective nonstationary nonresonancy can be attained against the earthquake ground motions that have various characteristics.

Two types of active response control system are introduced in this paper. In an actual design, the selection of the most appropriate devices and systems depends on the type of external disturbance, the type of subject structure and the aim of control. In addition, energy saving, economics and simplification of the system would be considered and developed to achieve a final purpose in the practical system.

<References>
1. T.Kobori et al., 'Study on Active Mass Driver (AMD) System -Active Seismic Response Controlled Structure- (Part 1), (Part 2)', 4th World Congress of Council on Tall Buildings and Urban Habitat, Nov. 1990, Hong Kong

2. T.Kobori et al., 'Shaking Table Experiment of Multi-Story Seismic Response Controlled Structure with Active Variable Stiffness (AVS) System', The 8th Japan Earthquake Engineering Simposium, Dec. 1990

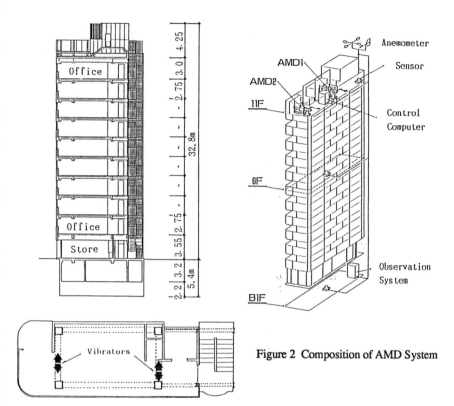

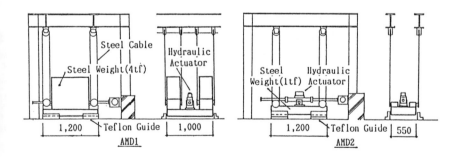

Office

Office

Store

Vibrators

Figure 1 Plan of Typical Floor and Cross Section

Anemometer

AMD1

AMD2

Sensor

11F

Control
Computer

6F

Observation
System

BIF

Figure 2 Composition of AMD System

Steel Cable

Steel Weight(4tf)

Hydraulic
Actuator

1,200   Teflon Guide   1,000

AMD1

Steel
Weight(1tf)

Hydraulic
Actuator

1,200   Teflon Guide   550

AMD2

Figure 3 Details of the Driving Units   (unit:mm)

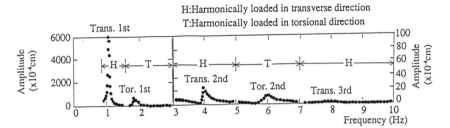

Figure 4 Resonance Curve of Displacement with AMD off (10th floor)

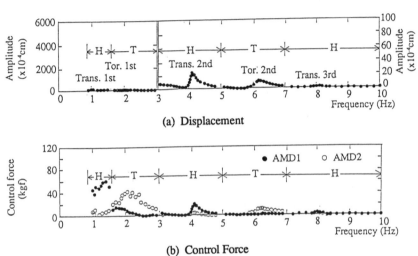

(a) Displacement

(b) Control Force

Figure 5 Resonance Curve with AMD on (Test)

Table 1 Resonance Period and Damping Factor

| Direction | Mode No. | Vibration test | | | Analytical model |
		Resonance period (s)	Resonance frequency (Hz)	Damping factor (%)*	Resonance period (s)
Transverse	1	0.94	1.065	0.77	0.93
	2	0.25	3.98	1.35	0.27
	3	0.12	8.00	2.26	0.13
Torsional	1	0.54	1.85	1.96	0.54
	2	0.17	5.99	2.43	0.17

* Evaluated from free vibration test

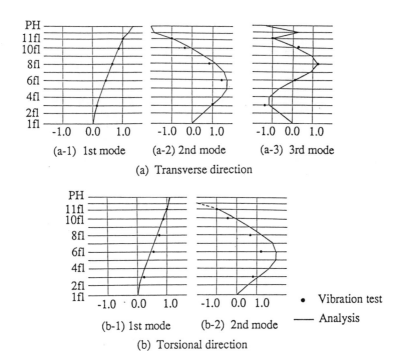

(a-1) 1st mode (a-2) 2nd mode (a-3) 3rd mode

(a) Transverse direction

(b-1) 1st mode (b-2) 2nd mode

• Vibration test
—— Analysis

(b) Torsional direction

Figure 6 Natural Vibration Modes

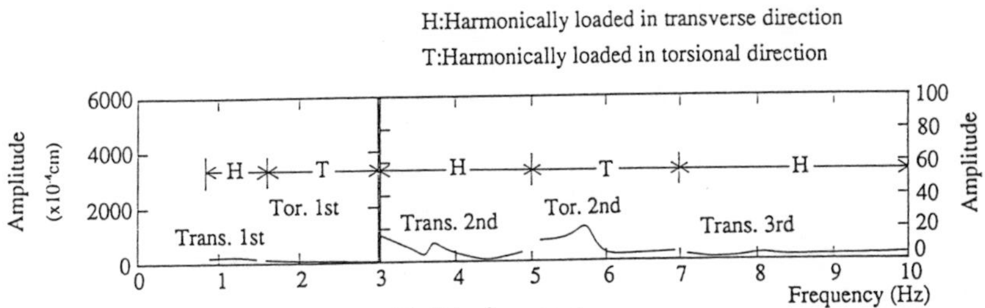

H:Harmonically loaded in transverse direction
T:Harmonically loaded in torsional direction

(a) Displacement

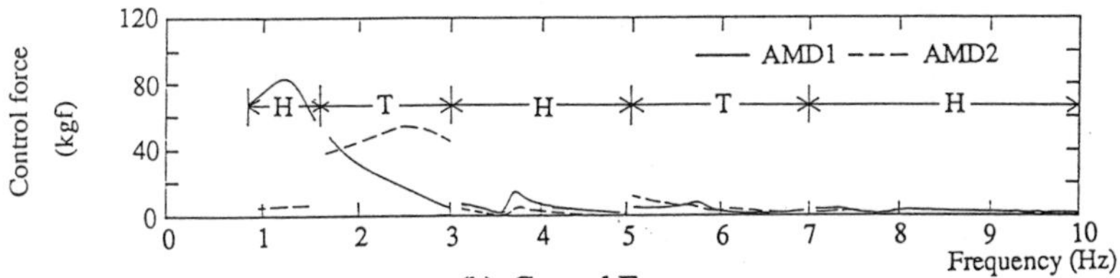

(b) Control Force

Figure 7 Resonance Curve with AMD on (Analysis)

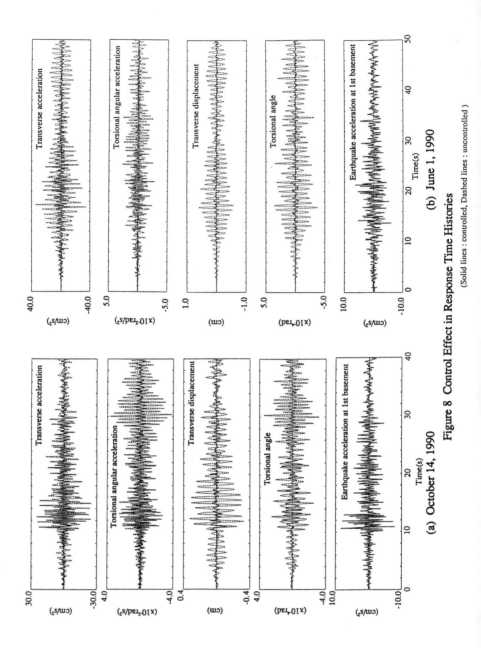

(a) October 14, 1990

(b) June 1, 1990

Figure 8 Control Effect in Response Time Histories

(Solid lines : controlled, Dashed lines : uncontrolled)

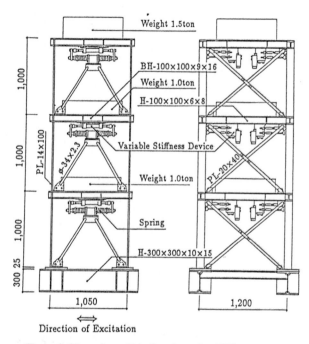

Weight 1.5ton

BH-100×100×9×16
Weight 1.0ton
H-100×100×6×8

Variable Stiffness Device

Weight 1.0ton

Spring

H-300×300×10×15

PL-14×100

α-34×2.3

PL-20×40

1,000
1,000
1,000
300 25

1,050   1,200

⟸⟹
Direction of Excitation

Figure 9 Three Story Test Specimen for AVS system

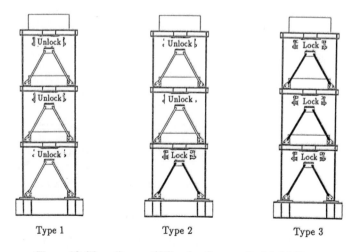

Type 1   Type 2   Type 3

Figure 10 Three Types of Vibration Systems for Model Test

p 161

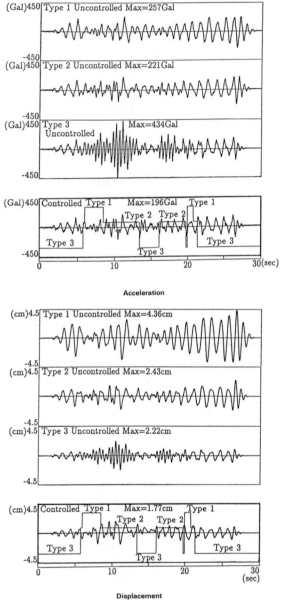

Figure 11 Top Floor Responses under Mexico Earthquake

p 162

SENSITIVITY ANALYSIS AND IDENTIFICATION ON BASE-ISOLATED STRUCTURAL SYSTEM

Chin-Hsiung Loh[*]

Chin-Ching Kuo[**]

ABSTRACT

The response of uncertain base isolation system to stochastic excitation was studied in the first part. This base isolation system is idealized as a laminated rubber bearings with linear viscous damper and linear spring. It was found that the most sensitive parameters to this type of system are the mass ratio (between super-structure and basement), natural frequency and damping of isolators. Finally, the recorded earthquake response of a base-isolation building — the Law and Justice Center in Rancho Cucamonga, shaken by the 1990 Upland earthquake (M_L = 5.5) is discussed by system identification techniques. It is point out that the isolator filtered out most of the input high frequency seismic waves.

INTRODUCTION

Seismic isolation is a design strategy based on the premise that it is possible to uncouple the structure from the ground and thereby protect it from the damaging effects of earthquake motion. To achieve this additional flexibility isolation system is introduced at the base of the structure. A mechanism of additional damping is also provided so as to control the deflection which across the isolation interface. In recent years, use of base-isolation systems for protecting structures against earthquakes has attracted considerable attention. Excellent survey articles on historical development and recent literature on this subject were provided by Kelly.[1,2] The most commonly used base-isolation system is the laminated rubber bearing (LRB). The LRB consists of alternating layers of rubber and steel with the rubber being vulcanized to the steel plates. The mechanical behavior of this system is a simple model with linear spring and linear viscous damping. Since most isolation system are intrinsically non-linear, and the effective stiffness and damping will have to be estimated by some equivalent linearization process.[3,4,5]

The main purpose of this study contains two parts. The first part is the response analysis of an uncertain base-isolation system to stochastic excitation. In addition to the reliability analysis of the system, the sensitivity measures to system parameters is also examined. With the ideas of these reliability studies, the second part of this study is to perform the system identification of base-isolated structure during earthquake. The San Bernardino County Law and Justice Center in Rancho Cucamonga during the February 28 (1990) earthquake near Upland are used for this identification. The dynamic properties of the structure and the isolation system during earthquake are studied.

RESPONSE OF AN UNCERTAIN BASE-ISOLATION SYSTEM TO STOCHASTIC EXCITATION

The first section is to study the reasons of an uncertain base-isolation system to earthquake by the method of reliability analysis. The primary focus is one the effect of uncertainty of modal parameters in both the isolation system and super-structural system on the reliability. The set of random variables used to model the isolation system and the response are transformed into the space of uncorrelated normal variates, and first- and second-order approximations are used to estimate the reliability. In addition to the reliability, the sensitivity of the reliability to system parameters is examined.

A linear multi-story shear type structure is considered mounted on a base-isolated foundation. This base isolation system provides stiffness and viscous type damping in the horizontal direction. The equation of motion of this isolation system (shown in Figure 1) for ground excitation are

$$[M]\,\{\ddot{x}\} + [C]\,\{\dot{x}\} + [K]\,\{x\} = -[M]\,\{1\}\,(\ddot{x}_g + \ddot{x}_b)$$

$$m_b\,(\ddot{x}_g + \ddot{x}_b) + c_b\,\dot{x}_b + k_b\,x_b - c_2\,\dot{x}_2 - k_2\,x_2 = 0 \tag{1}$$

[*] Professor, Department of Civil Engineering, National Taiwan University, Taipei, Taiwan, R. O. C.

[**] Research Assistant, Department of Civil Engineering, National Taiwan University, Taipei, Taiwan, R. O. C.

in which $[M]$, $[C]$, and $[K]$ are the mass, damping, and stiffness matrices of the super-structure. Adopting normal mode approach and assuming that the super-structure vibrates in the first mode, one may derive the following system for the base and modal displacement:[6]

$$\ddot{x}_b + 2\xi_b \omega_b \dot{x}_b + \omega_b^2 x_b - 2\xi_1 \omega_1 d\mu \dot{y} - \omega_1^2 y d\mu = -\ddot{x}_g$$

$$a\,\ddot{x}_b + \ddot{y} + 2\xi_1 \omega_1 \dot{y} + \omega_1^2 y = -a\,\ddot{x}_g \tag{2}$$

in which a is the participation factor, y is the dimensionless modal displacement, and $\mu = m_N/m_b$, $\omega_b = \sqrt{k_b/m_b}$. The complex frequency response functions for y and x_b corresponding to $\ddot{x}_g = e^{i\omega t}$ can be calculated, i.e., $x_b = H_b(\omega)\,\ddot{x}_g$ and $y = H_1(\omega)\,\ddot{x}_g$. In this study, the modified Clough-Penzien spectrum was adopted as power spectral density of the input ground acceleration:

$$G_{\ddot{x}_j}(f) = \frac{1 + 4\xi_g^2(f/f_g)^2}{\left[1-(f/f_g)^2\right]^2 + \left[2\xi_g\,(f/f_g)\right]^2} \cdot \frac{(f/\bar{f})^4}{\left[1-(f/\bar{f})^2\right]^2 + \left[2\xi_1\,(f/\bar{f})\right]^2} \cdot \frac{f^2}{f^2+A^2}\,\overline{G}_0 \tag{3}$$

where $\overline{G}_0$ is the constant, ξ_g, f_g, f_1, $\bar{f}$, and A are the parameters. Figure 2 shows the normalized power spectral density of this input spectrum.

Numerical Results of the Isolation System: The reliability of a one story base-isolated LRB system with uncertain properties is examined. The stationary Gaussian base acceleration with a given frequency spectrum shown in Equation (3) was used as input excitation and the system equation is shown in Eq. (2) with $a = 1.0$. The stationary output displacement spectrum for the basement or super-structure is given by

$$G_x(\omega|x_1,\cdots,x_5) = \left|H(\omega|x_1,x_2,x_3,x_4,x_5)\right|^2\,G_{\ddot{x}_0}(\omega) \tag{4}$$

in which $H(\omega|\cdot)$ depends on the natural frequency (x_1) and damping ratio (x_2) of the super-structure; the natural frequency (x_3) and viscous damping ratio (x_4) of the isolation system; and the uncertain mass ratio between super-structure and the basement, x_5. For the present example, the input ground motion parameters are modeled as deterministic parameters. All the uncertainty parameters are assumed to be statistically independent random variables with normal distributions. For reliability analysis, the failure criteria for either structural response or basement response is expressed as

$$G(\vec{x}) = x_7 - x_6 \leq 0 \tag{5}$$

in which x_7 is the allowable deformation and x_6 is the relative displacement between the super-structure and the foundation, or the relative displacement between the basement and the foundation. $\vec{x}$ is the vector which contains the parameters $x_1 \sim x_7$. Because the assumption of independence between parameters $(x_1 \sim x_5)$, the inverse transformation leads to

$$G(\vec{x}) = x_7 - F_{x_6}^{-1}[\Phi(u_6)|x_1,\cdots,x_5] \leq 0 \tag{6}$$

in which $F_{x_6}^{-1}$ is an inverse with respect to the distribution of displacement x_6. For the purpose of this example, the distribution developed by Vanmarcke,[7] is used:

$$F(x_6|x_1,\cdots,x_5) = \left[1 - \exp\left(\frac{-x_6}{2\lambda_0}\right)\right] \exp\left[-\nu\tau\,\frac{1 - \exp\left(-\sqrt{\frac{\pi}{2}}\,q\,\frac{x_6}{\sqrt{\lambda_0}}\right)}{\exp\left(\frac{x_6^2}{2\lambda_0}\right) - 1}\right] \tag{7}$$

which in terms of the spectral moment, equivalent duration (τ) of strong motion and the dispersion measurement q. The HL-RF method is used to approximate the probability of a system exceeding a specified threshold. The major problem in applying this method is to calculated the gradient vector of failure criteria in the standard normal space in terms of the gradient vector in the original space:

$$\nabla_u\,G(\vec{u}) = J_{u,x}^{-1}\,\nabla_x\,G(\vec{x}) \tag{8}$$

p 164

in which $g(\vec{u})$ is a function in the space of standard normal, and $J_{u,x}$ is the Jacobian matrix. In the numerical analysis of the base-isolation system, we use the relative displacement between the basement and the foundation to serve as the failure criteria. Figure 3a and 3b show the reliability index of this system by considering ω_b and ξ_b and mass ratio m_1/m_b as random variables with different degree of uncertainty (variation of coefficient of variation in system parameter). Figure 3b considered the effect of rocking to the isolation system. In this study we choose the most sensitive variates, i.e. ξ_b, ω_b and mass ratio, to the uncertainty analysis.

Sensitivity Measures: It is often of interest to know the sensitivity of the reliability index to variations in parameters in the safety problem. To measure the total change in reliability when uncertainty in the system parameter is included, the following sensitivity measure is defined:[8]

$$D_i = \frac{\nabla G_i^2}{\nabla G_{i+1}^2} \qquad \text{for} \quad i = 1, 2, \cdots, n \qquad (9)$$

where ∇G_i is the ith coordinate of gradient vector evaluated at $\vec{u}_0$. Figure 4 shows the sensitivity measures of D_i. In general, the most sensitive measures influencing reliability are as follows: the natural frequency of the isolator ω_b, the damping ratio of the isolator ξ_b, and the mass ratio.

IDENTIFICATION ON A BASE-ISOLATED BUILDING DURING EARTHQUAKE

The Law and Justice building in Rancho Cucamonga is a 4-story braced steel-frame structure isolated on elastomeric bearings located between the basement and foundation levels. The building is instrumented with a total of 16 accelerometers which are located at the foundation, basement, 2nd floor and roof levels. An earthquake of magnitude 5.5 (M_L) occurred on February 28, 1990 at a distance of about 12 km from the building. The peak horizontal acceleration was 138.9 cm/sec^2 at the foundation level (below the isolator), 52.6 cm/sec^2 at the basement level (above the isolators), and 152.8 cm/sec^2 at the roof level. Preliminary Fourier analysis of the data in NS direction indicates that the high frequency signals were filtered out as waves propagate from foundation to basement, as shown in Figure 5. The structure above the isolators behaved as a 1.5 Hz for fundamental frequency.

Based on the proposed Equation (2), the equation of motion for super-structure can be rearranged in the following form:

$$\ddot{x}_t + 2\xi_1\omega_1\dot{x}_t + \omega_1^2 x_t = (1-a)\,\ddot{x}_0 + 2\xi_1\omega_1\dot{x}_0 + \omega_1^2 x_0 \qquad (10)$$

in which $\ddot{y} + \ddot{x}_b + \ddot{x}_g = \ddot{x}_t$ and $\ddot{x}_b + \ddot{x}_g = \ddot{x}_0$. Since $\ddot{x}_t$ and $\ddot{x}_0$ can be measured directly as the roof and basement acceleration response, the frequency response of acceleration input $(\ddot{x}_0)$ and acceleration output $(\ddot{x}_t)$ is expressed as

$$H(\omega|f_1, \xi_1, a) = \frac{(1-a) - 2\xi_1(f_1/f)\,i - f_1^2/f^2}{(1 - f_1^2/f^2) - 2\xi_1(f_1/f)\,i} \qquad (11)$$

Equation (11) gives the frequency response function by considering basement acceleration as input and roof acceleration as output. Since $\ddot{x}_0(t)$ and $\ddot{x}_t(t)$ are the measures signals, the best estimates of this frequency response function is through the following equation

$$\tilde{H}(\omega) = S_{\ddot{x}_0\ddot{x}_t}(\omega) \Big/ S_{\ddot{x}_0\ddot{x}_0}(\omega) \qquad (12)$$

in which $S_{\ddot{x}_0\ddot{x}_t}(\omega)$ is the cross spectrum between input and output. The modal parameters can be estimated by minimizing the following function

$$J = \int_{f_0}^{f_1} \left| \tilde{H}(\omega) - H(\omega|f_1, \xi_1, a) \right|^2 d\omega \qquad (13)$$

Once the structural parameters are estimated, the isolator parameters can also be estimated by using the first equation of Equation (2) to estimate ξ_b, ω_b and $d\mu$ from frequency domain analysis by assuming linear behavior of the isolators.

p 165

Figure 6 shows the comparison of the estimated as well as the calculated frequency response function. The estimated structural natural frequency is about 1.45 Hz and the natural frequency of the isolator is about 2.5 Hz. Although the comparison in frequency response function is not quite close, especially at low frequency of 0.5 Hz. More detail study on the dynamic behavior of this isolated structure during this earthquake is necessary.

CONCLUSIONS

The purpose of this study is to study the reliability of an uncertain base isolation system, subjected to stochastic excitation. The isolation system was assumed to have linear viscous damping and stiffness characteristics. Numerical results show that the uncertainty of modal parameters (based on the modal analysis) can be expressed as a relationship between the reliability index and the coefficient of variation of modal variables. It is also found that the mass ratio of super-structure to basement, natural frequency and damping ratio of isolators are the most sensitive parameter to this system. Through system identification technique on the data of Upland earthquake, Law and Justice Center in Rancho Cucamonga, the dynamic characteristics of isolator and the building are identified. Assuming linear behavior on the building as well as the isolator (equivalent linear), the natural frequency and damping ratio of super-structure are 1.45 Hz and 0.10, respectively. The identified equivalent natural frequency and damping ratio of isolator are 2.4 Hz and 0.28, respectively for N-S direction (3.0 Hz and 0.22, respectively for E-W direction). From Fourier analysis of the recorded data, it is clear that the isolator filter out most of the high frequency signals of seismic input.

ACKNOWLEDGMENTS

The work described in this paper was supported in part by the National Science Council of R. O. C. through grants NSC80-0414-P002-16B is appreciated.

REFERENCES

1. Kelly, J. M., "Aseismic Base Isolation," Shock and Vibration Digest, 1982, 14, pp. 17–25.

2. Kelly, J. M., "Aseismic Base Isolation: Review and Bibliography," Soil Dyn. Earthquake Engineering, 1986, 5, pp. 202–216.

3. Su, Lin, G. Ahmadi and I. G. Tadjbakhsh, "Responses of Base-Isolated Shear Beam Structures to Random Excitations," Probabilistic Engineering, Mechanics, Vol. 5, No. 1, 1990, pp. 35–46.

4. Su, Lin, G. Ahmadi and I. G. Tadjbakhsk, "A Comparative Study of Performances of Various Base Isolation Systems — Part I: Shear Beam Structure," Earthquake Engineering Struct. Dyn., 1989, 18, pp. 11–32.

5. Su, Lin, G. Ahmadi, and I. G. Tadjbakhsk, "A Comparative Study of Performances of Various Base-Isolation Systems — Part II: Sensitivity Analysis," Earthquake Engineering, Struct. Dyn., 1990, 19, pp. 21–33.

6. Constantinou, M. and I. G. Tadjbakhsk, "Probabilistic Optimum Base Isolation of Structures," J. of Structural Engineering, ASCE, Vol. 109, No. 3, 1983, pp. 676–689.

7. Vanmarcke, E. H., "Structural Response to Earthquakes," Seismic Risk and Engineering Decisions (L. Lomnitz and E. Rosenblueth editors), Elservier Book Co., The Netherlands, 1976.

8. Igusa, T. and Derkiureghian, A., "Response of Uncertain Systems to Stochastic Excitations," J. of Engineering Mechanics, Vol. 114, No. 5, May 1988, pp. 812–832.

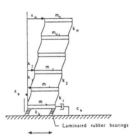

Fig.1: Model of basic system.

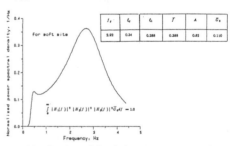

Fig.2: Normalized input power spectra.

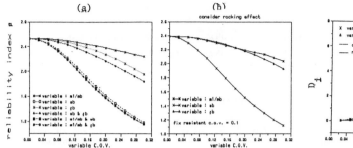

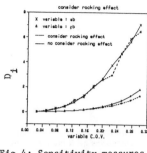

Fig,3: Plot of reliability index as a function of coefficient of variation of modal parameters. (a) without considering the rocking effect, (b) consider the effect of rocking.

Fig.4: Sensitivity measures of Di for the system parameters.

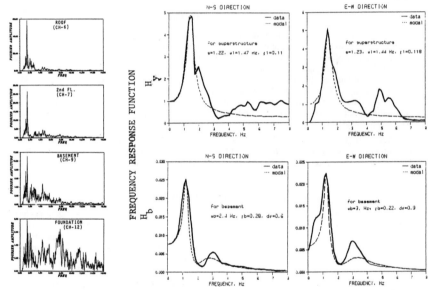

Fig,5: Fourier amplitude spectrum of recorded motion at different levels (N-S direction).

Fig,6: Comparison between the calculated & estimated frequency response function for superstructure and basement.

CONTROL OF THE SEISMIC RESPONSE OF STRUCTURES BY USE
OF BASE ISOLATION AND ACTIVE ABSORBING BOUNDARIES

J. E. Luco[1], H. L. Wong[2] and A. Mita[3]

ABSTRACT

The relative advantages of several control strategies to reduce the seismic response of multi-storey structures are studied. The strategies involve the separate or combined use of passive base isolation mechanisms and active control forces. The base isolation mechanism is modelled as an equivalent linear soft storey with high damping. The active control forces are selected so that an absorbing boundary is obtained at the top of the structure and a non-reflecting boundary is obtained at the base of the building. It is found that the best results are obtained when a passive base isolation system is combined with an active absorbing boundary placed at the top of the building. However, the incremental gains resulting from adding a base isolation system to a structure already controlled by a roof-top active absorbing boundary are significant only for relatively soft base isolation systems. Also, the incremental gains appear to decrease as the number of storeys of the structure increases.

INTRODUCTION

In this paper we examine the possible advantages of combining a base isolation system with active control in the form of absorbing and/or non-reflecting boundaries. For the purpose of the study, the base isolation system is represented by an equivalent linear soft-storey and the superstructure is represented by a uniform lumped mass model (Fig. 1). The active control system is represented by forces acting at the top and base of the structure. The control force at the top is selected to result in an absorbing boundary at the roof level. This absorbing boundary is such that all waves travelling towards the top of the structure are absorbed at the top and no downward reflected waves are obtained. The base force is selected to result in a non-reflecting boundary at the base level such that no energy is trapped within the base isolation system.

The control strategy used in this work is based on concepts of wave propagation or power flow (Vaughan, 1968; von Flotow, 1986; Mita and Luco, 1990a,b; Luco et al, 1990). This strategy differs from the typical base isolation approaches (Kelly, 1986) and from the usual active control systems (Yang and Soong, 1988).

BASIC EQUATIONS

We consider harmonic vibrations of the structural model shown in Fig. 1. The vibrations have time dependence $e^{i\omega t}$, where ω is the frequency. The model consists of N equal storeys of mass m_1 and complex stiffness $\tilde{K}_1 = k_1[1 + 2i\xi_1\text{sign}(\omega)]$ where k_1 is the stiffness and ξ_1 is the structural damping ratio. The floor diaphragms are considered to be rigid and the columns are assumed to be inextensible. When a base isolation system is included it is represented by the mass m_0 and by the equivalent complex stiffness $\tilde{K}_0 = k_0[1 + 2i\xi_0\text{sign}(\omega)]$ where k_0 is the equivalent stiffness and ξ_0 the equivalent structural damping ratio. Control forces $F_N e^{i\omega t}$ and $F_0 e^{i\omega t}$ may be applied to the roof and base level. The structure is subjected to a horizontal ground motion represented by u_g and the effects of soil-structure interaction are not included in the analysis. The equations of motion of the structure in this case are given by

$$-\omega^2 m_1 u_N + \tilde{K}_1(u_N - u_{N-1}) = F_N \tag{1}$$

$$-\omega^2 m_1 u_j + \tilde{K}_1(2u_j - u_{j+1} - u_{j-1}) = 0, \quad (1 \leq j \leq N - 1) \tag{2}$$

$$-\omega^2 m_0 u_0 + \tilde{K}_0(u_0 - u_g) + \tilde{K}_1(u_0 - u_1) = F_0 \tag{3}$$

[1] Department of Applied Mechanics and Engineering Sciences, University of California, San Diego, La Jolla, Ca 92093-0411.
[2] Department of Civil Engineering, University of Southern California, Los Angeles, California 90089-2531.
[3] Ohsaki Research Institute, Shimizu Corporation, Fukoku Seimei, Bldg. 2-2-2, Uchisaiwai-cho, Chiyoda-ku, Tokyo, 100, Japan.

where $u_j e^{i\omega t}$ is the absolute displacement of the j-th level. The control forces at the top (absorbing boundary) and base (non-reflecting boundary) are given by [Luco et al (1990)]

$$F_N = -\dot{u}_N m_1 \tilde{\omega}_1 e^{-i\gamma_1/2} a_l \tag{4}$$

$$F_0 = -\left[m_0 \tilde{\omega}_0 e^{-i\gamma_0/2} - m_1 \tilde{\omega}_1 e^{-i\gamma_1/2} \right] \dot{u}_0 b_l \tag{5}$$

where $\dot{u}_N = i\omega u_N$, $\dot{u}_0 = i\omega u_0$, and a_l, b_l take values of 0 or 1 depending on the case considered. In Eqs. (4) and (5)

$$\gamma_p = 2\sin^{-1}\left(\frac{\omega}{2\tilde{\omega}_p}\right) \ , \ \tilde{\omega}_p = \left(\frac{\tilde{K}_p}{m_p}\right)^{1/2} = (1 + 2i\xi_p)^{1/2}\sqrt{\frac{k_p}{m_p}} = (1 + 2i\xi_p)^{1/2}\omega_p \ , \ (p = 0, 1) \ . \tag{6}$$

It should be noted that if $\xi_1 = 0$, then γ_1 is real for $\omega < 2\omega_1$ and complex for $\omega > 2\omega_1$. If $\xi_1 \neq 0$, then γ_1 is complex for all frequencies. The corresponding behavior of γ_0 depends on whether $\omega \lessgtr \omega_0$.

Six cases are considered. These cases correspond to: (1) a reference case with no base isolation and no control forces [$a_1 = b_1 = 0$, $k_0 = \infty$], (2) active absorbing boundary at top of structure and no base isolation [$a_2 = 1$, $b_2 = 0$, $k_0 = \infty$], (3) passive base isolation only [$a_3 = b_3 = 0$], (4) active absorbing boundary at the top of the building and passive base isolation at the base [$a_4 = 1$, $b_4 = 0$], (5) active absorbing boundary at top, active non-reflecting boundary at the base and base isolation [$a_5 = 1$, $b_5 = 1$] and (6) active non-reflecting boundary at the base and base isolation [$a_6 = 0$, $b_6 = 1$].

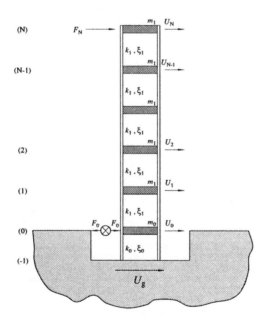

Fig. 1. Model of Structure Consisting of N Equal Storeys of Mass m_1, Stiffness k_1 and Structural Damping Ratio ξ_1. The Base Isolation Mechanism is Represented by the Equivalent Stiffness k_0 and Equivalent Damping Ratio ξ_0. Active Control Forces F_0 and F_N are Applied at the Base and at the Roof Levels. The Ground Motion is Represented by U_g.

The solution to Eqs. (1), (2) and (3) can be obtained by standard methods [Thomson (1988), p. 341] in which u_j is written in the form

$$u_j = A e^{-i\gamma_1 j} + B e^{i\gamma_1 j} \quad , \quad (j = 0, N) \tag{7}$$

where the constants A and B can be found by imposing Eqs. (1) and (3). The first and second terms on the right-hand-side of Eq. (7) can be considered to represent upgoing and downgoing waves, respectively.

If an absorbing boundary is applied at the top then $B = 0$ and $A = u_0$ needs to be obtained from Eq. (3). If no absorbing boundary is applied at the top then both A and B need to be determined by Eqs. (1) and (3). The solution for all six cases ($l = 1$ through 6) can be written in the form [Luco et al (1990)]

$$u_j = \begin{cases} u_g \cos[\gamma_1(N - j + \tfrac{1}{2})]/\Delta_l & , \quad l = 1, 3, 6 \\ u_g e^{-i\gamma_1 j} & , \quad l = 2 \quad (j = 0, N) \\ u_g \exp(-i(\gamma_0 + \gamma_1 j))/\Delta_l & , \quad l = 4, 5 \end{cases} \tag{8}$$

where

$$\Delta_1 = \cos[\gamma_1(N + \tfrac{1}{2})] \tag{9a}$$

$$\Delta_3 = \cos[\gamma_0 + (N + \tfrac{1}{2})\gamma_1] + 2\sin\left(\frac{\gamma_0}{2}\right)\left\{\sin[(N + \tfrac{1}{2})\gamma_1 - \tfrac{1}{2}\gamma_0] - \frac{m_1}{m_0}\frac{\tilde{\omega}_1}{\tilde{\omega}_0}\sin(N\gamma_1)\right\} \tag{9b}$$

$$\Delta_6 = \Delta_3 + 2i\sin\left(\frac{\gamma_0}{2}\right)\cos\left[\gamma_1\left(N + \tfrac{1}{2}\right)\right]\left[e^{-i\gamma_0/2} - \frac{m_1}{m_0}\frac{\tilde{\omega}_1}{\tilde{\omega}_0}e^{-i\gamma_1/2}\right] \tag{9c}$$

$$\Delta_4 = 1 + 2e^{-i\gamma_0}\sin\left(\frac{\gamma_0}{2}\right)\left\{\left(\frac{m_1}{m_0} - 1\right)\sin\left(\frac{\gamma_0}{2}\right) + i\left[\frac{m_1}{m_0}\frac{\tilde{\omega}_1}{\tilde{\omega}_0}\cos\left(\frac{\gamma_1}{2}\right) - \cos\left(\frac{\gamma_0}{2}\right)\right]\right\} \tag{9d}$$

and

$$\Delta_5 = 1 \tag{9e}$$

Eqs. (8) and (9) indicate that resonance of the structure is eliminated when the absorbing boundary is applied at the top of the structure (Cases $l = 2$, 4 and 5). In the absence of the absorbing boundary at the top (Cases $l = 1$, 3 and 6) resonant behavior can be obtained at some characteristic frequencies.

NUMERICAL RESULTS

As an example we consider a ten-storey structure ($N = 10$, $f_1 = \omega_1/2\pi = 6.37$ Hz, $\xi_1 = 0.05$) and characterize the base isolation system (when used) by $f_0 = 3.18$ Hz, $m_0 = m_1$ and $\xi_0 = 0.20$. In this example, the equivalent stiffness k_0 of the base isolation system is 1/4 of the stiffness k_1 of the other floors, i.e., $k_0/k_1 = (\omega_0/\omega_1)^2(m_0/m_1) = 1/4$. The amplitudes of the calculated transfer functions $U_T(\omega) = u_N/u_g$, $U_B(\omega) = u_0/u_g$, $F_T(\omega) = F_N/m_1\omega_1\dot{u}_g$, and $F_B(\omega) = F_0/m_1\omega_1\dot{u}_g$ for the six cases described in the previous section are shown versus frequency $f = \omega/2\pi$ in Fig. 2. In Case 1, corresponding to the structure without base isolation and with no control forces, the transfer function $U_T(\omega)$ for the displacement at the top shows an amplification of the order of 12 at the fundamental frequency of the structure. If a control force in the form of an absorbing boundary is applied at the top of the structure (Case 2) then resonance is eliminated and the amplitude of the transfer function $U_T(\omega)$ is less than 1 for all frequencies. In Cases 1 and 2 the motion of the base u_B is equal to the ground motion u_g. When a passive base isolation system (Case 3) is used without control forces on the roof or base, the transfer function $U_T(\omega)$ shows peak amplification of the order of 5 at a modified fundamental frequency of about 0.75 Hz. For frequencies above 3 Hz, the base isolation system drastically reduces the response at the top and base of the structure. It is apparent that the passive base isolation system is very effective in curtailing the high-frequency response but that it does not eliminate the resonance at low frequencies.

Combination of a passive base isolation system with an absorbing boundary at the top (Case 4) results in the complete elimination of resonance and in absolute displacements at the top and base of the structure significantly lower than the ground motion u_g. As shown by the transfer function F_T, the control force at the top required in Case 4 is significantly lower than that in Case 2 when no base isolation system is used.

p 170

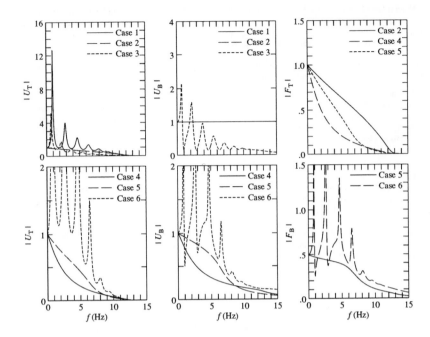

Fig. 2. Amplitudes of the Transfer Functions $U_T(\omega)$, $U_B(\omega)$, $F_T(\omega)$ and $F_B(\omega)$ ($N = 10$, $f_0 = 3.18$ Hz, $\xi_0 = 0.20$, $f_1 = 6.37$ Hz, $\xi_1 = 0.05$, $m_0/m_1 = 1$).

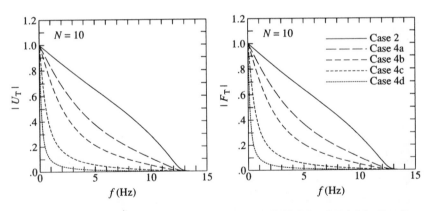

Fig. 3. Comparison of the Amplitudes of the Transfer Functions $U_T(\omega)$ and $F_T(\omega)$ for Case 2 (Absorbing Boundary at Top Only) with those for Case 4 (Absorbing Boundary at Top Combined with Passive BaseIsolation) for (a) $f_0 = 4.5$, (b) 3.18, (c) 1.59 and (d) 0.795 Hz. Results for a $N = 10$ Storey Structure are Presented for $f_1 = 6.37$ Hz, $\xi_1 = 0.05$, $\xi_0 = 0.20$ and $m_0 = m_1$.

p 171

When a base isolation system is combined with a control force at the top and base of the structure as in Case 5 resonance is also eliminated and values of $|U_T(\omega)|$ and $|U_B(\omega)|$ less than one are obtained for all frequencies. The amplitudes of $U_T(\omega)$, $U_B(\omega)$ and $F_T(\omega)$ for Case 5 and $0 < f < 7$ Hz are somewhat higher than the corresponding values for Case 4. For this reason, Case 4 is considered preferable to Case 5 since it leads to a lower response (at least up to 7 Hz) and does not require a control force at the base. Finally, the results for Case 6 corresponding to a base isolation system with a control force at the base but no control force at the top show amplification of the ground motion at both the top and base of the structure by a factor larger than 2.

The amplitudes of the transfer functions $U_N(\omega) = u_N/u_g$ and $F_T(\omega) = F_N/(m_1\omega_1\dot{u}_g)$ for Case 2 are compared in Fig. 3 with those for Case 4 for four values of the equivalent stiffness of the base isolation system. Clearly, a softer base isolation system leads to a lower response at the top and to a lower required control force.

CONCLUSIONS

We have studied the relative advantages of various strategies to control the seismic response of structures. These strategies involve the combined or separate use of base isolation and active control forces. As in previous studies (Mita and Luco, 1990a, b), we have found that the use of a control force to obtain an absorbing boundary at the top of the building results in the complete elimination of any resonant behavior in the structure. The required control force, however, can be large; its peak value can be of the order of the weight of one floor of the building. We have also found that a passive base isolation system (represented in this study by an equivalent soft and highly damped first story) while highly effective in reducing the high-frequency response does not affect the low frequency response to the same degree. For very tall buildings the low-frequency response is not reduced unless the base isolation system is made extremely soft.

When a passive base isolation system is used in combination with an active absorbing boundary at the top of the building then the response at all frequencies is controlled to be lower than the amplitudes of the input ground motion. Also, the resulting control force is somewhat lower than the force required when no base isolation is used. By combining a sufficiently soft base isolation system with active control by absorbing boundaries a synergistic effect is obtained in which some of the drawbacks of base isolation are eliminated while at the same time lower control forces are required thus facilitating the implementation of the active control.

A control system based on the combined use of an active absorbing boundary at the top of the building together with an active non-reflecting boundary at the base and a base isolation mechanism also eliminates all resonant behavior in the structure. However, the response in this case is larger than the response obtained when base isolation is combined with an absorbing boundary at the top without any control force at the base.

ACKNOWLEDGEMENT

The work described here was supported by grants from the Ohsaki Research Institute, Shimizu Corporation to the University of California, San Diego and the University of Southern California.

REFERENCES

Kelly, J. M. (1986)."Aseismic Base Isolation: Review and Bibliography," Soil Dynamics and Earthquake Engineering, Vol. 5, No. 3, 202-216.

Luco, J. E., H. L. Wong and A. Mita (1990). "Active Control of the Seismic Response of Structures by Combined Use of Base Isolation and Absorbing Boundaries," Earthquake Engineering and Structural Dynamics, (submitted for publication).

Mita, A. and J. E. Luco (1990a). "New Active Vibration Control Strategy for Tall Buildings," Proceedings Eighth Japan Earthquake Engineering Symposium (submitted for publication).

Mita, A. and J. E. Luco (1990b). "Active Vibration Control of a Shear Beam with Variable Cross Section," Proceedings of the 1990 Dynamics and Design Conference, Japan Society of Mechanical Engineers (Kawazaki, Japan, July 9-12), 276-279.

Thomson, W. T. (1988). Theory of Vibrations with Applications, (Third Edition), Prentice-Hall, Englewood Cliffs, New Jersey.

Vaughan, D. R. (1968). "Application of Distributed Parameter Concept to Dynamic Analysis and Control of Bending Vibrations," Journal of Basic Engineering, June, 157-166.

Yang, J. N. and T. T. Soong (1988). "Recent Advances in Active Control of Civil Engineering Structures," Probabilistic Engineering Mechanics, Vol. 3, 179-188.

On Rendering Structures Earthquake-Resistant by
Retrofitting with Active Controls

Leonard Meirovitch

Department of Engineering Science and Mechanics
Virginia Polytechnic Institute and State University
Blacksburg, VA 24061

Abstract

This paper presents an approach for retrofitting structures so as to withstand future earthquakes. The approach consists of structural modeling, parameter identification and design of active controls, all carried out in a consistent fashion.

1. Introduction

During earthquakes structures are often subjected to violent loads, exceeding the intensity for which they were designed. On occasions, the result is catastrophic failure. On other occasions, although catastrophic failure does not occur, the structure is weakened to the extent that its viability comes into question. In particular, the question arises as to whether the structure is able to withstand future earthquakes. The concern is that such structures will collapse partially or totally. Even if there is no danger of collapse, displacements during earthquakes can be of such magnitude so as to cause discomfort or damage to nonstructural members and/or property. In such cases, it is necessary to determine the state of the structure so as to permit a decision whether the existing damage can be mitigated by retrofitting the structure to bring its response to a safe level, or whether the damage is so severe that the structure must be condemned.

Before the retrofitting of a structure so as to withstand future earthquakes can be undertaken, it is necessary to determine the current state of the structure. This implies determination of the excitation - response characteristics of the structure. Such characteristics should enable one to detect regions in which the structural integrity has been impaired. The implication is that visual inspection alone is not able to reveal these regions, as the regions may not be exposed to the naked eye. Hence, the interest lies in undertaking a testing program in conjunction with a certain synthesis procedure based on measurements from these tests and capable of predicting the response characteristics of the structure. The program consists of exciting the structure by means of given forces and/or torques and measuring the response at a number of judiciously chosen points. Then, the synthesis consists of inferring the structural parameters from the excitation and response of the structure. The process of synthesizing the parameters from the excitation and response is known as structural parameter identification.

Before a parameter identification can be undertaken, it is necessary to postulate a mathematical model for the structure. To this end, one can consider the finite element method (Ref. 1) or a Rayleigh-Ritz based substructure synthesis (Ref. 2). In both cases the parameters, regarded as unknown, assume the character of the discretization process. In the finite element model, the parameters play the role of local unknowns associated with each finite element (Ref. 3). On the other hand, in substructure synthesis the parameters are expressed as series of given admissible functions multiplied by unknown coefficients, as in Ref. 4. It is clear that the nature of the parameter identification process must be consistent with the type of model adopted.

Assuming that a suitable mathematical model has been constructed and the structural parameters have been identified, the question remains as to how to improve the response characteristics of the structure. In some cases, the damage is not severe, so that use of passive control is sufficient. This implies the use of damping and stiffening materials. In other cases, the nature of the damage requires active control.

The design of controls must be consistent with the type of model used. We distinguish between two control procedures, direct output feedback control and modal control. The first uses collocated sensors and actuators and the actuator force at a given location depends only on the sensor output at the same location (Ref. 5). Such a control scheme is more suitable for a finite element model. A second technique, known as the independent modal-space control method (Ref. 5), is more suitable for use in conjunction with the Rayleigh-Ritz based substructure synthesis (Ref. 2).

2. The Equations of Motion for Parameter Identification
    Structures are basically distributed-parameter systems, which are governed by partial differential equations. However, except for some very simple structures, a formulation in terms of partial differential equations is likely to cause extreme difficulties. For example, the frame shown in Fig. 1 consists of six structural members connected at four joints, in addition to two support points. Ignoring axial vibration, each of the structural members can be described by one partial differential equation for the bending vibration and two boundary conditions at each end point. The boundary conditions relate the motions in any two adjacent members, so that the problem is very complex. In fact, it is so complex that no closed-form solution is possible. As a result, spatial discretization is a virtual necessity, which amounts to replacing the partial differential equations by sets of simultaneous ordinary differential equations. As mentioned in the Introduction, spatial discretization can be carried out by the finite element method or by a substructure synthesis based on the classical Rayleigh-Ritz method. Our interest is not so much in the derivation of the equations of motion as in putting them in a form suitable for parameter identification.

Let us refer to the structure of Fig. 1 and consider spatial discretization by means of the finite element method. Figure 1 shows a typical finite element i with the nodal displacements and nodal forces; m_i, c_i and k_i denote the mass, damping and stiffness distributions over the element. Following the usual procedure (Ref. 1), the equations of motion can be written in the compact matrix form

$$M\ddot{\underaccent{\tilde}{w}}(t) + C\dot{\underaccent{\tilde}{w}}(t) + K\underaccent{\tilde}{w}(t) = \underaccent{\tilde}{F}(t) \qquad (1)$$

where $\underaccent{\tilde}{w}(t)$ is an n-dimensional nodal displacement vector, M, C and K are $n \times n$ mass, damping and stiffness matrices, respectively, and $\underaccent{\tilde}{F}(t)$ is an n-dimensional nodal force vector. Introducing the notation

$$L = [M \; C \; K], \quad \underaccent{\tilde}{x}(t) = \left[\ddot{\underaccent{\tilde}{w}}^T(t) \; \dot{\underaccent{\tilde}{w}}^T(t) \; \underaccent{\tilde}{w}^T(t) \right]^T \qquad (2a,b)$$

Eq. (1) can be rewritten as

$$L\underaccent{\tilde}{x}(t) = \underaccent{\tilde}{F}(t) \qquad (3)$$

Equation (3) forms the basis for the parameter identification algorithm to follow.

p 174

3. A Parameter Identification Algorithm

The object is to uncover regions of weakness in a structure. Such regions are characterized by reduced stiffness compared to expected stiffness. For example, in a structure like the frame of Fig. 1 the beams and columns can be expected to have uniform stiffness. Hence, if in a certain region the stiffness exhibits an unexpected drop, then this region can be regarded as one of impaired strength.

The parameters appear explicitly in the matrices M, C and K. It is very common to base parameter identification algorithms on the identification of the matrices M, C and K. This procedure, however, is not recommended here. Indeed, even taking into consideration that the matrices M, C and K are symmetric, the number of entries in these matrices is considerably larger than the number of parameters. The reason for this is that the same parameter tends to appear in a number of entries, so that the entries are not independent quantities. Hence, our interest is in a parameter identification algorithm in which the independent parameters are placed in evidence.

With reference to Fig. 1, let us introduce the notation

$$p_r = \begin{cases} m_r, & 1 \le r \le N \\ c_{r-N}, & N+1 \le r \le 2N \\ k_{r-2N}, & 2N+1 \le r \le 3N \end{cases} \tag{4}$$

where m_r, c_{r-N} and k_{r-2N} are the distributed mass, damping and stiffness over element r, respectively, in which N is the number of finite elements. Then, the matrix L defined by Eq. (2a) can be expressed as

$$L = \sum_{r=1}^{3N} \frac{\partial L}{\partial p_r} p_r \tag{5}$$

This permits us to rewrite Eq. (3) in the form

$$A(t)\underset{\sim}{p} = \underset{\sim}{B}F(t) \tag{6}$$

where now the system parameters appear as unknowns, in which

$$A(t) = [\underset{\sim}{a_1}(t)\, \underset{\sim}{a_2}(t) \cdots \underset{\sim}{a_{3N}}(t)]$$
$$= \left[\frac{\partial L}{\partial p_1} \underset{\sim}{x}(t)\ \frac{\partial L}{\partial p_2} \underset{\sim}{x}(t) \cdots \frac{\partial L}{\partial p_{3N}} \underset{\sim}{x}(t) \right] \tag{7}$$

is an $n \times 3N$ time-dependent matrix of coefficients and $\underset{\sim}{p} = [p_1\, p_2 \cdots p_{3N}]^T$ is a 3N-dimensional vector of parameters.

Next, let us sample the matrix $A(t)$ and the vector $\underset{\sim}{F}(t)$ at the times $t = t_k$ $(k = 1,2,...,m)$ and introduce the notation

$$B = \begin{bmatrix} A(t_1) \\ A(t_2) \\ \cdot \\ \cdot \\ \cdot \\ A(t_m) \end{bmatrix}, \quad c = \begin{bmatrix} \underline{F}(t_1) \\ \underline{F}(t_2) \\ \cdot \\ \cdot \\ \cdot \\ \underline{F}(t_m) \end{bmatrix} \tag{8a,b}$$

Then, the equations of motion, Eq. (6), sampled at the indicated times can be expressed as

$$D\underline{p} = \underline{c} \tag{9}$$

which represents a set of $m \cdot n$ equations and $3N$ unknowns, where m must be such that $m \cdot n \geq 3N$. The solution of Eq. (9) can be obtained by the least-squares method and can be written as (Ref. 3)

$$\underline{p} = D^\dagger \underline{c} \tag{10}$$

where

$$D^\dagger = (D^T D)^{-1} D^T \tag{11}$$

is the pseudo-inverse of D.

The bottom third of the vector p is of particular interest as it represents the stiffness distribution of the structure in a finite element sense, i.e., a sectionally-constant distribution, where each section corresponds to one finite element. If the vector contains a sudden drop in value, which cannot be explained by a shift from one structural member to another, then the drop must be attributed to a weakening of the structure.

4. Retrofitting with Active Controls

We are interested in the case in which the parameter identification process described in the preceding section reveals regions of impaired strength. These local weaknesses can be the result of earthquake damage, but it can be from other sources as well, such as poor construction or poor design. We propose here to improve the structure response characteristics by means of active control. It is common in civil engineering applications to think in terms of tendon control. Due to its nature, tendon control must be part of the initial design and construction, which rules it out for retrofitting. In fact, retrofitting is best carried out be means of point sensors as actuators. As sensors, one can consider accelerometers and rate gyros and as actuators proof-mass actuators, air jet thrusters and control moment gyros. These are point sensors and actuators more commonly encountered in aerospace applications than in civil structures. Of course, their advantage in retrofitting is that they can be installed without disturbing the structure unduly.

In the case of earthquakes, the absolute displacements can be written as

$$\underline{w} = T\underline{u}_s + \underline{u} \tag{12}$$

p 176

where $T\underset{\sim}{u}_s$ is a vector of rigid-body nodal displacements due to the motion of the supports, in which T is a transformation matrix, and $\underset{\sim}{u}$ is a vector of elastic deformations. Then, inserting Eq. (12) into Eq. (11), the equations of motion can be written in the form

$$M\underset{\sim}{\ddot{u}} + C\underset{\sim}{\dot{u}} + K\underset{\sim}{u} = B\underset{\sim}{F} + \underset{\sim}{F}_s \qquad (13)$$

where

$$\underset{\sim}{F}_s = -MT\underset{\sim}{\ddot{u}}_s - CT\underset{\sim}{\dot{u}}_s - KT\underset{\sim}{u}_s \qquad (14)$$

is an inertial force vector due to the motion of the support, which can be regarded as a persistent disturbance.

The control can be divided into a closed-loop part designed to control transient disturbances and another part designed to control persistent disturbances, where the latter can be open loop. Hence, we express the control vector as follows:

$$\underset{\sim}{F} = \underset{\sim}{F}_c + \underset{\sim}{F}_o \qquad (15)$$

where the open-loop control is such that

$$B\underset{\sim}{F}_o = -\underset{\sim}{F}_s \qquad (16)$$

so that the open-loop control is given by

$$\underset{\sim}{F}_o = -B^{\dagger}\underset{\sim}{F}_s \qquad (17)$$

in which $B^{\dagger}$ is the pseudo-inverse of B, obtained by replacing D by B in Eq. (11)

For the closed-loop control, we propose to use direct output feedback control (Ref. 5), whereby pairs of collocated sensors and actuators are located at judiciously chosen nodal points throughout the structure. The control forces satisfy the control law

$$F_{ci}(t) = -g_i u_i(t) - h_i \dot{u}_i(t), \quad i = 1,2,...,r \qquad (18)$$

where g_i and h_i are control gains. Equations (18) can be written in the vector form

$$\underset{\sim}{F}_c(t) = -GB^T\underset{\sim}{u}(t) - HB^T\underset{\sim}{\dot{u}}(t) \qquad (19)$$

Hence, combining Eqs. (13), (17) and (19), we obtain the closed-loop equation

$$M\underset{\sim}{\ddot{u}}(t) + (C + BHB^T)\underset{\sim}{\dot{u}}(t) + (K + BGB^T)\underset{\sim}{u}(t) = (I - BB^{\dagger})\underset{\sim}{F}_s \qquad (20)$$

The gain matrices G and H are chosen so that the components of the deformation vector $\underset{\sim}{u}(t)$ decay as fast as possible without reaching large amplitudes. The positive definiteness of the closed-loop coefficient matrices $C + BHB^T$ and $K + BGB^T$ guarantees good stability characteristics.

5.   Summary and Conclusions
This paper outlines an approach whereby structures impaired by earthquakes, poor design or poor construction can be retrofitted so as to withstand future earthquakes. The approach consist of three phases, structural modeling, parameter identification and control design, all compatible with one another.

p 177

The modeling must be such as to permit ready parameter identification and control design. As far as parameter identification is concerned, the object is to identify physical parameter distributions, rather than abstract coefficient matrices. This affords greater versatility in the event the modeling must be refined. Finally, the number and location of the sensors and actuators, as well as the control gains, must be such as to ensure maximum stiffening of the structure and optimal damping possible. In this regard, it should be pointed out that the stiffening restores structural integrity and damping mitigates overall damage to property and enhances comfort during earthquakes.

6. References

1. Meirovitch, L., Computational Methods in Structural Dynamics, Sijthoff and Noordhoff, The Netherlands, 1980.

2. Meirovitch, L. and Kwak, M. K., "On the Modeling of Flexible Multi-Body Systems by a Rayleigh-Ritz Based Substructure Synthesis," AIAA Journal (to appear).

3. Meirovitch, L. and Norris, M. A., "On the Problem of Modeling for Parameter Identification in Distributed Structures," International Journal for Numerical Methods in Engineering, Vol. 28, 1989, pp. 2451-2463.

4. Baruh, H. and Meirovitch, L., "Parameter Identification In Distributed Systems," Journal of Sound and Vibration, Vol. 101, No. 4, 1985, pp. 551-564.

5. Meirovitch, L., Dynamics and Control of Structures, Wiley-Interscience, New York, 1990.

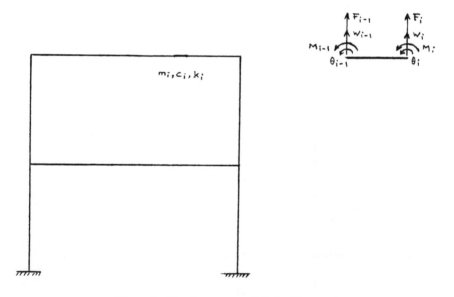

Figure 1. The Structure and Finite Element i

Current Research Activities of Shimizu for Response Control

Yorihiko Ohsaki
Executive Vice President, Shimizu Corporation
Professor Emeritus, University of Tokyo
2-2-2 Uchisaiwai-cho, Chiyoda-ku, Tokyo 100, Japan

ABSTRACT

An overview of the past and present research activities of Shimizu Corporation in the field of structural control is presented. A number of passive and active control systems have been studied and developed for wind loads and/or earthquake loads. Some of these passive control systems have been already implemented in several actual structures. The passive control systems summarized here include base-isolation systems, visco-elastic dampers, tuned liquid dampers, tuned mass dampers and others. For the active control of structures we have recently developed a hybrid mass damper which combines a tuned mass damper with electric actuators. This combination allows us to control a structure with considerably less power than typical active mass dampers. The passive control effects of the tuned mass damper help to reduce the control force and the required power. In addition, searching for ways of reducing the control energy, a mechanism of active control of the energy flow for earthquake inputs and wind loads is now being investigated. New control algorithms resulting from this study based on the propagation of waves within the structure are briefly explained and the possibility of applying these algorithms in conjunction with the hybrid mass damper is discussed.

INTRODUCTION

Even in this era of high technology, it is still true that simplicity and reliability should be the most important features of structural control devices. In fact, most of the sophisticated active control devices do nothing but to add structural damping and/or to shift the natural frequencies of the structure. If it is possible to achieve the same effects with passive control devices, the need for other active alternatives is reduced. Consequently, Shimizu Corp. has devoted major efforts to the development of a number of passive devices, such as base isolation systems, visco-elastic dampers, tuned mass dampers and tuned liquid dampers. Some of these devices have been already used in several actual buildings.

The use of active control systems to reduce the response of buildings during large earthquakes presents many difficulties. We started the research by evaluating the amount of energy required to control the structure. It turned out that, except for small buildings, active control devices used against earthquake inputs involve high energy requirements. Even for wind loads a fully active device requires large control forces. For this reason, we decided to develop hybrid systems which combine passive and active devices. As a first step, we have recently developed a hybrid mass damper (HMD) which combines a tuned mass damper with electric actuators. The HMD drastically reduces the control force and energy requirement while keeping the same performance than a fully active system for wind loads.

The search for possible ways to reduce the control energy requirements for both earthquake and wind loads has led to the study of the active control of the energy flow within structures. From this point of view, we have reached a simple approach to response control, that is, controlling boundary conditions rather than modes of vibration. Two types of control methods, an absorbing boundary and a reflecting boundary, have been proposed. The absorbing boundary placed at the top of a building can suppress all components of modes. The reflecting boundary is also powerful when combined with a soft first story, e.g.: a base isolation system. A shaking table experiment was conducted recently to verify this new concept. Some aspects of this experiment will also be presented.

PASSIVE CONTROL SYSTEMS

The simplicity and reliability of passive vibration control systems have motivated much work during the past several years. Many passive control systems have been studied and developed for wind loads and/or earthquake loads for a variety of structures. Most of these systems are aimed at augmenting structural damping in the target structures or shifting the natural frequencies. In the following, some of the passive control systems utilized by Shimizu Corp. are introduced. They include: base-isolation systems, visco-elastic dampers, tuned liquid dampers and tuned mass dampers. Even for structures with active control systems, some amount of damping augmentation will be essential considering the spillover problems. In that sense, development of more efficient passive control systems is a most important and urgent task for structural control.

An important passive control system for use against large earthquakes has been recently developed jointly with Prof. Akiyama at the University of Tokyo and the system has been recently applied to the Ultra Clean Room Laboratory built at the Institute of Technology, Shimizu Corp.[1] This system is designed to concentrate input earthquake energy into some portion of the target structure (e. g.: the first story) and to absorb it by an energy absorbing device. The concentration of the input energy is accomplished by designing the first story softer than the other stories. The concentrated energy is absorbed by hysteretic dampers. This design philosophy is closely related to a new active control algorithm explained later.

Base-Isolation System

Several types of base isolation systems have been studied analytically and experimentally using reduced models. Shimizu has also conducted full-scale tests at Sendai in collaboration with Tohoku University.[2, 3, 4] Two tests buildings, base-isolated and ordinary founded, were constructed side by side in a yard of Tohoku University in Sendai, about 200 miles north from Tokyo. The systems tested at the site include rubber bearing with viscous dampers and high viscosity rubber bearings. More than ten earthquake events have been recorded during a one-year-and-half period. The data obtained from this joint effort is available to any researchers in this field.

Figure 1. Full-Scale Building Tests for Base-Isolation Systems

Other types of isolation systems are currently being tested jointly with Argonne National Laboratory. A shaking table test using scaled models is now underway in collaboration with Prof. Kelly at UCB.

Visco-Elastic Dampers

As an important device which can be attached to most types of conventional buildings a visco-elastic damper consisting of rubber-asphalt has been developed.[5] This type of damper is very efficient to reduce the response especially in the high frequency range and at low vibration levels. A prototype of this damper system has been applied to our new office building currently under construction in the Shibaura area of Tokyo as shown in Fig. 2. Forced vibration tests will be conducted just before completion of construction by the end of this year followed by earthquake observation.

Figure 2. Visco-Elastic Dampers Installed in Office Building at Shibaura

Tuned Liquid Dampers

Tuned liquid dampers are simple and effective passive control devices especially for wind loads. The control force is generated due to the sloshing motion of the liquid in the container. Analytical tools for evaluating the linear as well as nonlinear motion have been developed so that systematic design for these dampers is currently available. Several sets of the tuned liquid dampers have been applied in several actual structures.[6] An example is shown in Fig. 3. This damper consists of plastic containers and water and is simple and easy to maintain. By adjusting the viscosity of the liquid a new type of the damper is currently under development to improve its efficiency.

Figure 3. Tuned Liquid Dampers Installed in Yokohama Marine Tower

Tuned Mass Dampers

A prototype of a tuned mass damper (TMD) has been developed which consists of layers of rubber bearings and a mass as shown in Fig. 4. This TMD has an air brake at its center for safety reasons. The model shown in Fig. 4 has been used for shaking tests.[7] This combination is excellent since only a small area of floor is required for installation. In addition, the vertical motion of the mass coupled with the horizontal motion is much smaller than for a wire-suspended system. Since the rubber bearing itself has been extensively used for base-isolation systems, the reliability of the current TMD system is excellent. Soon a TMD similar to the prototype in Fig. 4 is going to be installed in a symbolic tower at Nagasaki.

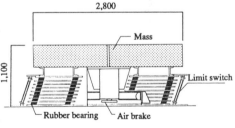

Figure 4. Tuned Mass Damper for Validation Tests

ACTIVE CONTROL SYSTEMS

In most cases, even the most sophisticated active control systems simply add structural damping or shift vibration frequencies. Because of their cost, energy requirements and reliability considerations, passive alternatives are far better especially for large earthquakes if such devices are available. However, for wind loads, an active mass damper, for example, has several advantages compared to a tuned mass damper since the active device can reduce torsional vibrations and be effective for a wider frequency range. The active system also has a superior capability to reduce the vibration amplitudes. But again, the energy requirements for a tall building for which such device would be applicable are still large. This is why we have recently developed a hybrid mass damper (HMD) which combines the tuned mass damper with electric actuators. This particular combination allows us to reduce the control energy by seventy to eighty percent compared to a typical active mass damper for wind loads. A brief description of the HMD is given below.

For large seismic loads, many difficulties need to be overcome to apply active control systems to tall buildings against large earthquakes. The possibility of reducing the control energy requirement has led to the study of the active control of the energy flow for earthquake inputs. From this point of view, we have reached simple approach to response control based on controlling boundary conditions rather than vibration modes. Two types of control methods, an absorbing boundary and a reflecting boundary, have been proposed. This new control algorithm is presented for a shear structure model. A shaking table test using a six-story model has been conducted and some data validating the new concept are explained below. The possibility of employing this algorithm in conjunction with the HMD is explored and the efficiency of this approach is compared with the conventional LQ control algorithm.

Hybrid Mass Damper (HMD)

A description of the HMD is given in Fig. 5. This system consists of the tuned mass damper shown in Fig. 4 and electric actuators located under the mass. The control force is applied to the mass at its center so that no torsional motion is generated. This HMD can control the translational modes in two horizontal directions. The actuators consist of AC servo motors and ball screws. The specification of this HMD is depicted in Table 1. This system was designed for a small building which has a long natural period. However, the maximum control force of this system is rather large compared to the weight of the mass so that the current system can be used not only for wind loads but also for medium seismic loads. If the HMD is designed only for wind loads, the required control force should be smaller.

A shaking table test has been conducted recently and the capability of the HMD was confirmed. This system is planed to be mounted on an actual building and verification tests, wind and earthquake observations will be conducted.

Table 1 Specification of HMD

TMD specification
$m_d g$ = 4.0ton
k_d = 0.0745 ton/cm
T_d = 1.47sec
h_d = 0.03

Actuator specification
F_{max} = 1.2ton
P_{max} = 7.5kw

Figure 5.   Hybrid Mass Damper

Boundary Condition Control Systems
Considering the waves propagating along the structure, two new active control algorithms, the absorbing boundary and the reflecting boundary, have been recently proposed.[8), 9)] Similar control algorithms have been studied by others for very flexible space structures modeled as simple continuous systems. While this approximation may be appropriate for some flexible structures, it may not be adequate for many buildings. Although the

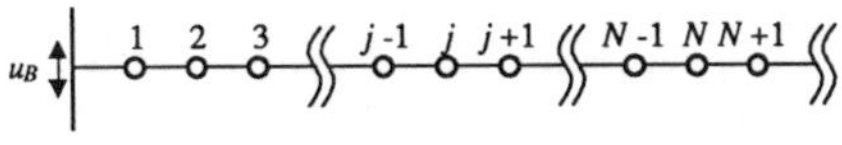

Figure 6. Infinitely Tall Building Consisting of Equal Masses and Springs

discrete structure model considered here is quite simple, it is expected that it will bring forth the key differences between continuous and discrete systems.

For an infinitely tall building (Fig. 6) consisting of equal masses m and springs k, the equation of motion of the j-th mass is given by

$$- \omega^2 m u_j + k\,(2u_j - u_{j-1} - u_{j+1}) = 0 \quad , \quad (j = 1, 2, 3, ...) \tag{1}$$

where $u_j e^{i\omega t}$ is the absolute displacement of the j-th mass. If the motion of the base is u_B then the response at the j-th mass is described by

$$u_j = u_B\,e^{-i\gamma j} \tag{2}$$

in which γ is related to the wave number of a uniform shear bar and is given by

$$\gamma = 2\sin^{-1}\,(\omega/2\omega_0) \tag{3}$$

where $\omega_0^2 = k/m$. From Eq.(3) it is found that γ is a real number when $\omega < 2\omega_0$ and that γ becomes complex when $\omega > 2\omega_0$. In the absence of structural damping, Eq.(2) indicates that the motion applied at the base of the building propagates along the building with no reduction of its amplitude if the frequency is lower than the cut off frequency $2\omega_0$. For frequencies above $2\omega_0$ there is a very strong reduction in amplitude which at high frequencies is proportional to $(\omega_0/\omega)^{2j}$.

The force F_j that the portion of the building above the j-th mass exerts on that mass is given by

$$F_j = -\,m\omega_0 v_j e^{-i\,\gamma/2} \tag{4}$$

where $v_j = \dot{u}_j = i\omega u_j$ is the absolute velocity of the j-th mass. Equation (4) indicates that for $\omega < 2\omega_0$ and in the absence of structural damping, the internal force F_j propagates along the structure without a change in amplitude. For $\omega > 2\omega_0$, the internal force F_j decreases in amplitude with height.

Absorbing Boundary Control Algorithm
It is easy to show that if the control force

$$F_N = -\,m\omega_0 v_N e^{-i\,\gamma/2} \tag{5}$$

is applied to the top of an N-story building then the response of the building when subjected to the base motion u_B will be identical to that of an infinitely tall building. The resulting motion u_j and internal force F_j are given, in this case, by Eqs.(2) and (4), respectively.

A comparison of the amplitudes of the transfer functions (u_N/u_B) for 100-story buildings with and without the energy absorbing boundary (A. B.) is presented in Fig. 7 for a damping ratio of $h=0.01$. Clearly, the absorbing boundary results in a major reduction of the response.

Since the control force given by Eq. (5) is frequency dependent, it may be necessary to use an approximate control force of the form

$$F_N = - K \, m \, \omega_0 \, v_N \, e^{-i \omega \delta} \qquad (6)$$

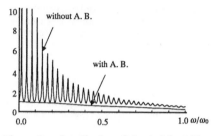

Figure 7. Amplitudes of (u_N/u_B) for 100-Story Building with and without Absorbing Boundary

where K is the force amplitude factor and δ is the time delay. The response of the N-story building for this control force is expressed by

$$u_j = u_B \frac{cos\left[\gamma\left(N - j + \frac{1}{2}\right)\right] + i \, K \, e^{-i \omega \delta} sin\left[\gamma(N - j)\right]}{cos\left[\gamma\left(N + \frac{1}{2}\right)\right] + i \, K \, e^{-i \omega \delta} sin(\gamma N)} \quad , \quad (j = 0, 1, 2, \ldots, N) \qquad (7)$$

The amplitude of the transfer function u_{10}/u_B for an ideal absorbing boundary is compared with that for an approximate absorbing boundary with time delay δ set at $(2\omega_0)^{-1}$ in Fig. 8. It is found that this approximation is appropriate at all frequencies. Two other examples for $\delta=0$ and $\delta=1/\omega_0$ are also plotted in Fig. 8. It is clear that the large time delay induces a large amplification in the high frequency range.

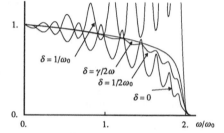

Figure 8. Effects of Time Delay on the Amplitude of (u_{10}/u_B) for 10-Story Building ($h=0.01$)

The effects of structural damping on the response of structures with absorbing boundaries are explored in Fig. 9. The responses at the top of 10- and 100-story buildings were calculated for $h=0$, 0.01 and 0.05. From Fig. 6 it is apparent that consideration of the structural damping reduces the amplitude of the response considerably. This reduction is more significant for the 100-story building.

The control force required to achieve an absorbing boundary at the top of an undamped building is independent of the height of the building but can be quite large. For a peak ground velocity of 25 cm/sec, the peak of the required control force would be of the order of the weight of one floor.

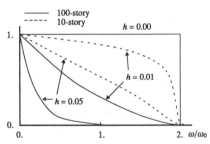

Figure 9. Effects of Structural Damping on the Ratio of (u_N/u_B) for 10- and 100-Story Buildings

Reflecting Boundary Control Algorithm

As an alternative, a reflecting boundary at which all incoming waves are totally reflected can be introduced at the M-th floor. In this case, and for the simple shear model under consideration, the resulting response is

$$u_j = \begin{cases} u_B \sin[\gamma(M-j)]/\sin(\gamma M) & , \quad j \leq M \\ 0 & , \quad j \geq M \end{cases} \tag{8}$$

The control force required at the M-th floor is given by

$$F_M = -k \, u_B \sin \gamma / \sin(\gamma M) \tag{9}$$

In the undamped case, both the control force and the response in the lower $(M-1)$ floors become very large at certain characteristic frequencies. However, this control algorithm may be efficient when combined with soft first stories such as base-isolation systems.

Experimental Validation

To validate the proposed algorithms, shaking table tests were conducted.[10), 11)] The model used and the disposition of sensors are shown in Fig.10. The stiffness of each column and the weight of each story are almost equally distributed. Therefore, this frame is suitable for confirming the effectiveness of the absorbing boundary. An electromagnetic exciter was used as a control force generator.

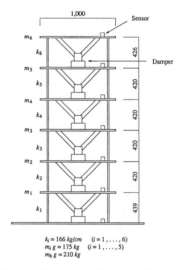

Since the cut-off frequency $2\omega_0$ of this frame is about 61rad/s (9.7Hz), the optimum control time delay is about 16ms. The damping ratio of the frame is 0.04 at the first natural frequency of 1.12Hz and it increases to 0.12 at the second natural frequency of 3.67Hz. For this particular model, the control force is given in the feedback form.

$$F_6 = -5.54 \, v_6 e^{-i\omega\delta} \tag{10}$$

$k_i = 166 \, kg/cm \quad (i = 1, \ldots, 6)$
$m_i g = 175 \, kg \quad (i = 1, \ldots, 5)$
$m_6 g = 210 \, kg$

Two sets of experiments, sinusoidal sweep experiments and earthquake excitation experiments (scaled El Centro 1940 NS), were conducted to examine the characteristics of the current system in the frequency domain and the time domain, respectively.

Figure 10. Scaled Six-Story Steel Frame Model

The experimental results of three cases, (a) without absorbing boundary, (b) with absorbing boundary (time delay $\delta=10$ms) and (c) with absorbing boundary ($\delta=65$ms), are compared in Figs.11(a)-(c) with simulated results. In these figures, some discrepancies between experimental and simulated results are observed mainly because the nonlinearity of dampers attached to the structure was not considered in the simulation. From Fig.11(b), it is found that the amplitude of the transfer function stays under the value of 1.0 for all frequency ranges and the phase varies almost linearly. This fact indicates that the waves coming from the shaking table were thoroughly absorbed at the top of the model. In Fig.11(c), the amplitudes of the experimental results are larger than those of the simulated results. This is due to the existence of control spillover primarily because of the characteristics of the exciter and, also, to the longer than appropriate time delay used in that case.

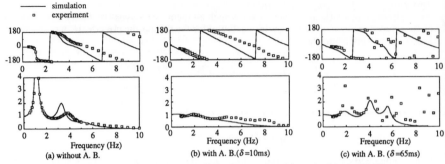

Figure 11. Frequency Response of the Structure with and without HMD

For the three cases described above, shaking table tests were carried out considering scaled El Centro 1940 NS inputs. The maximum base acceleration was about 20Gals. The peak acceleration at each floor of the model for the tests are plotted in Fig.12 after being normalized by the peak acceleration of the shaking table. From Fig. 12, it is clear that the absorbing boundary control with appropriate time delay suppresses resonance in all vibration modes and that the response at the upper floors is smaller than that at the lower floors. This decrease can be explained as the effect of structural damping.

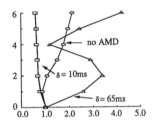

Figure 12. Normalized Peak Acceleration Response for El Centro 1940 NS

Absorbing Boundary Control Algorithm for HMD

In the tests just described, the absorbing boundary was established by an electromagnetic exciter which acts as an active mass damper. Active mass dampers usually have no spring or a very weak spring to be able to control all of the modes of interest. However, the hybrid mass damper can achieve the same absorbing boundary with less control force and energy.[12] The control force F_a which should be supplied by the actuator can be written in the form

$$F_a = - K \, m \, \omega_0 \, v_N \, e^{-i \, \omega \delta} - k_d(u_d - u_N) - c_d(v_d - v_N) \qquad (11)$$

in which d indicates values associated with the HMD. If this control force is realized, then the force establishing the absorbing boundary given by Eq. (6) is indeed applied to the structure.

A tall building model, which is modeled by 5 equal masses and springs for simplicity, is used for simulation. The weight of the mass is 1.0ton so that the total weight of the building is 5.0ton. The natural period of the first mode is 4.0sec. The damping characteristics of the building model are determined by a Rayleigh damping with a value of 0.01 for the first and second modes, respectively. A hybrid mass damper (HMD), with a weight of 0.04ton and natural period of 4.0sec is attached to the top of the model.

The frequency response of the building with the HMD was calculated for a particular earthquake input. The corresponding response with an AMD which has the same mass as the

HMD without the spring and damper is compared with that for the HMD. The control law for the AMD is given by Eq. (6). Both systems have the same reduction effects on the earthquake response since the total force applied through the actuator, the spring and damper is exactly equal. Due to this fact, the response of the damper mass and the capability of both systems are equivalent. However, the required actuator force is different because of the passive effects contributed by the spring and damper of the HMD. In Fig. 13, the control force of the actuators for both systems are compared for a value of K=0.2. It is clearly shown that the amplitude of the control force for the HMD is drastically smaller than that for the AMD around the fundamental natural frequency of the building model. Of course, no passive contribution is expected in the high frequency range. However, the spring and damper have no deleterious effects in the high frequency range. From this observation, it is expected that the HMD always needs a smaller control force than the AMD which results in less control energy for any type of the external force. This passive contribution can be much more significant if the external force has its major frequency content around the fundamental natural frequency of the building.

To evaluate the performance of the absorbing boundary control algorithm, comparisons were made with the conventional LQ control algorithm. The LQ control design employed here is to minimize the cost function defined as:

$$ J = \int_0^T [\, v_m^2 + r F_m^2 \,]\, d\,t \qquad (12) $$

in which v_m and F_m represent the modal velocity and the modal control force, respectively. Since it is difficult to distribute many sensors through the building to identify all modes, it is often assumed that the contribution of the fundamental mode is dominant and that the higher frequency content can be neglected. Therefore, the current LQ control design for earthquake inputs is done only considering the first mode of the building. By changing the weight r for the LQ control and K for the absorbing boundary control, the relation between the maximum acceleration amplitude at the top of the building model described above and the maximum power required for the actuator is plotted in Fig. 14 for El Centro 1940 NS input.

Because the LQ control neglects the higher modes of the building model, the acceleration response tends to diverge for control gains larger than certain level. This phenomena is known as a spillover effect. However, such effects do not appear for the absorbing boundary control. This fact should be kept in mind in the design of a control algorithm for the HMD since, in most cases, the number of sensors is limited.

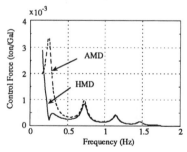

Figure 13. Actuator Force for HMD and AMD

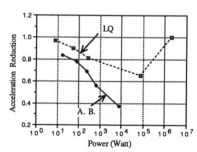

Figure 14. Response Reduction versus Power for Absorbing Boundary and LQ Control Algorithms

CONCLUDING REMARKS

An overview of the past and present research activities of Shimizu Corporation in the field of structural control has been presented. Among others, base-isolation systems, visco-elastic dampers, tuned liquid dampers and tuned mass dampers were briefly explained. Some of these passive control systems have been already applied in several actual structures.

For the active control of structures we have recently developed a hybrid mass damper which combines a tuned mass damper with electric actuators. This combination allows us to control a structure with considerably less power than a typical active mass damper since the passive control effects of the tuned mass damper help in reducing the control force and energy. In addition, a mechanism of active control of the energy flow within structures for earthquake and wind loads is now being investigated. New control algorithms based on the propagation of waves within the structure have been briefly explained and the possibility of applying the algorithms in conjunction with the HMD was discussed. Analytical studies have shown that the application of the absorbing boundary control algorithm to the HMD will result in the control of all vibration modes without inducing spillover problems and with less control energy than an active mass damper.

REFERENCES

1) Yabe, Y., S. Mase, H. Tsukagoshi, T. Hirama, T. Terada and F. Ohtake, "Pseudo-Dynamic Test on Full-Scale Model of First-Story Energy-Concentrated Steel Structure with Flexible-Stiff Mixed Frame," J. Struc. Constr. Engng., AIJ, No. 413, 41-52, 1990.

2) Tamura, K., H. Yamahara and M. Izumi, "Proof Tests of the Base-Isolated Buildings Using Full-Sized Model," Proc. Seismic, Shock, and Vibration Isolation-1988, PVP-Vol. 147, 21-28, 1988.

3) Kuroda, T., M. Saruta and Y. Nitta, "Verification Studies on Base Isolation Systems by Full-Scale Buildings," Proc. Seismic, Shock, and Vibration Isolation-1989, PVP-Vol. 181, 1-7, 1989.

4) Kaneko, M., K. Tamura, K. Maebayashi and M. Saruta, "Earthquake Response Characteristics of Base-Isolated Buildings," Proc. Fourth National Conference on Earthquake Engineering, Palm Springs, California, Vol. III, 569-578, 1990.

5) Nakamura, Y., K. Shiba, H. Yokota and M. Sukagawa, "Research on Vibration Control Methods (Part 1) Experiments of 6 Story Model with Rubber-Asphalt," Proc. AIJ Annual Meeting, Vol. B, 553-554, 1988.

6) Fujii, K., Y. Tamura, T. Sato and T. Wakahara, "Wind-Induced Vibration of Tower and Practical Application of Tuned Sloshing Damper," J. of Wind Engng., No. 37, 537-546, 1988.

7) Maebayashi, K., Y. Nakamura, K. Tamura, Y. Ogawa and H. Yamahara, "A Research on Hybrid Vibration Control System (Part 1) Shaking Table Test of TMD Using Multi-Stage Rubber Bearings," Proc. AIJ Annual Meeting, Vol. B, 787-788, 1990.

8) Mita, A. and J. E. Luco, "New Active Vibration Control Strategies for Tall Buildings," Proc. Eighth Japan Earthquake Engineering Symposium, in press.

9) Mita, A and J. E. Luco, "Active Vibration Control of a Shear Beam with Variable Cross Section," Proc. 1990 Dynamics and Design Conference, Kawasaki, 276-279, 1990.

10) Mita, A., K. Shiba, Y. Kitada, K. Naraoka and M. Kaneko, "Traveling Wave Control for Tall Buildings (Part 1 Theory)," Proc. AIJ Annual Meeting, Vol. B, 857-858, 1990.

11) Shiba, K., Y. Kitada, A. Mita, K. Naraoka and M. Kaneko, "Traveling Wave Control for Tall Buildings (Part 2 Experimental Study)," Proc. AIJ Annual Meeting, Vol. B, 859-860, 1990.

12) Kaneko, M. and A. Mita, "Active Vibration Control of Tall Buildings by a Hybrid Mass Damper," Proc. 1990 Dynamics and Design Conference, Kawasaki, 286-289, 1990.

ABSTRACT submitted to U.S. National Workshop on Structural Control Research

Analysis of Recorded Earthquake Response and
Identification of a Multi-story Structure Accounting
for Foundation Interaction Effects

A.S. Papageorgiou
Department of Civil Engineering, Rensselaer Polytechnic Institute, Troy, NY 12180-3590, USA.

Bing-Chang Lin
Department of Civil Engineering, Chung-Yuan University, Chungli, Taiwan 32023, Taiwan, People's Republic of China.

Analysis of recorded earthquake response and identification, of the 14-story, reinforced concrete Hollywood Storage building shaken during the 1987 Whittier Narrows earthquake, are presented. Since only the translational components of motion of the foundation were recorded and considered as input motions in the identification analysis, the inferred parameters reflect not only the characteristics of the superstructure but also the characteristics of the foundation and of the underlying soil, and as such are referred to as *apparent system parameters*. Further processing of the apparent system parameters is performed in order to extract the modal parameters of the superstructure. The pronounced interaction of the superstructure with the soil in the longitudinal direction reduced considerably (25%) the base shear which the building would have experienced if the superstructure were supported in a rigid half-space. On the other hand, the weak soil-structure interaction in the transverse direction reduced the base shear only by a very small amount (-1.2%).

ACTIVE TUNED MASS DAMPER OF BASE ISOLATED STRUCTURES

Jun-Ping Pu* and James M. Kelly**

Abstract

The efficiency of active tuned mass damper that was installed on the base of isolated structures under seismic excitation is investigated. Optimal parameters of the damper are examined. Optimal control strategy is adopted so as to reduce both the response of the structure and the stroke of the absorber. The mass of the damper needed is small, and the system is quite efficient through small control energy.

Passive Tuned Mass Damper

The efficiency and the optimal parameters for passive tuned mass dampers have been investigated by many investigators. The vibration absorbers are adopted mainly to absorb the energy closed to resonance and to mininize the steady-state response of structures.

For an absorber attached to a one degree-of-freedom structure, let

$\mu = m/M =$ mass ratio = absorber mass / main mass,

f = frequency ratio = absorber frequency / structural frequency,

g = forced frequency ratio = excitation frequency / structural frequency, and

ξ_a = damping ratio of absorber.

The amplitude of the main mass is a function of the four essential variables mu,f,g and ξ_a. A plot of the transfer function of the main mass as a function of the frequency ratio g for the definite absorber mass shows that all curves for various values of the damping ratio pass through two points. It is suggested by Den Hartog[3] that the optimal stiffness of the absorber is to make the ordinates of these two points equal. The optimal damping ratio of the absorber is the one that has the curve passes with a horizontal tangent through the highest of these two points.

For an absorber with mass ratio equals to 0.1, Den Hartog[3] suggested that the frequency ratio equals to 10/11, and the damping ratio is about 0.2. The optimal damping ratio for the absorber is also about 0.2 for multi-degree-of-freedom light-damped structures as investigated by Warburton et al[1,5,6].

---

* Professor, Department of Civil Engineering, Feng Chia University, Taiwan, R.O.C. Currently Visiting Scholar in the Department of Civil Engineering, University of California at Berkeley, CA.

**Professor, Department of Civil Engineering, University of California at Berkeley, CA.

The two largest peaks of the transfer function of structure are not necessarily have the same ordinate if the frequency ratio of absorber is chosen according to Den Hartog's suggestion. Since the shape of the transfer function depends on the frequency ratio of the absorber, one may adjust the frequency ratio of the absorber for the definite mass ratio and damping ratio so that the ordinates of the two largest peaks be equal to each other, thus the steady-state response of the structure can be minimized. If the mass ratio of the absorber is chosen to be 0.1, and the damping ratio be 0.2, then the optimal frequency ratio will be 0.9, as shown in Figure 1.

Active Tuned Mass Damper

The passive tuned mass damper is efficient in reducing the response of the structure subjected to harmonic or wind excitation. However, there has not been a general agreement about the effectiveness for seismic loads<4>.

The introduction of active control system can reduce both the transient response of the structure and the stroke of the absorber. Most of the active tuned mass dampers investigated were installed on the top floor of the building for suppressing primarily the first fundamental mode of the structure<2>. Although the relative response of the structure under seismic excitation was examined, only few of the investigations examined the absolute response.

The characteristics of base isolated structures differ from that of classical structures in that most of the high frequency responses were filtered out so that the dominant mode is almost like a rigid body motion. The shear force in the superstructure is small at the expense of quite large of the absolute response.

The active tuned mass damper of base isolated structure is not necessary to be installed on the top floor. In fact, it can be set on the base of the structure so that the seismic energy can be dissipated before it is transmitted to the superstructure. Besides, the installation is quite practical and easy.

Numerical Examples

For the sake of examining the effect of active mass damping for base isolated structure, the responses of three buildings with 1 story, 3 stories and 5 stories (including the base) were investigated. Tuned mass dampers were installed on the base. Assume that the buildings were subjected to El Centro earthquake (N-S May, 1940). The masses and stiffness of each floor of the building were assumed to be identical. The mass was chosen to be 252 Ton and the stiffness be $2.1 \times 10^5 \text{Ton/sec}^2$. The damping ratio of the structure was assumed to be 2%. The mass ratio of the absorber is chosen to be 0.1, and the damping ratio be 0.2.

Let the objective of the active control system is to minimize the following cost function:

$$J = \int_0^{t_f} [\sum_{i=a,1}^{n} (Q_1 x_i^2 + Q_2 \dot{x}_i^2 + Q_3 x_a^2 + Q_4 \dot{x}_a^2) + R u^2] \, dt$$

in which, x is the absolute displacement, $\dot{x}$ is the absolute velocity of the structure, x_a is the stroke of the absorber, u is the control force between base and tuned mass damper, and t_f is the final control time.

Let us examine and compare the following objectives: $J_1 = \int_0^{t_f} x_1^2 \, dx$, $J_2 = \int_0^{t_f} x_n^2 \, dx$,

$J_3 = \int_0^{t_f} x_a^2 \, dx$, $J_4 = \int_0^{t_f} \dot{x}_1^2 \, dx$, $J_5 = \int_0^{t_f} \dot{x}_n^2 \, dx$, $J_6 = \int_0^{t_f} \dot{x}_a^2 \, dx$, $J_7 = \int_0^{t_f} u^2 \, dx$,

in which x_1 and x_n are the absolute displacement of the base, and that of the uppermost floor respectively.

Some typical result was shown in Figure 2 to Figure 5. The maximum control forces needed is shown in Figure 6.

Conclusion

1. The optimum frequency ratio of passive TMD is near to 0.9 for a 1-story structure. As the building is higher, the optimum frequency ratio is shifted. Although the optimum parameters studied in the steady-state response can not be applied directly to the real state response, the difference is not much.

2. The response of the structure can be reduced to a large extend through passive TMD. The introduction of active control system will increase slightly the response when the stiffness of the passive TMD is near to its optimum value.

3. The response of the structure with active or passive TMD will be smaller than that without TMD only when the frequency ratio of the TMD is smaller than some specific value. That is, the stiffness of the TMD should not be too high especially for higher buildings.

4. The response of the structure with active TMD is almost independent to the frequency ratio of the TMD when the frequency ratio is small, say, less than 1.2. But the control force needed is smallest when the TMD has its optimum frequency ratio.

5. The efficiency of TMD is worse when the building is higher.

6. The maximum control force needed is very small, just about 4% of the weight of one floor.

7. The stroke of the absorber can be reduced dramatically if the active control system is introduced. The stroke of active TMD is almost independent to the frequency ratio of the absorber.

References

1. Ayorinde, E. O. and Warburton, G. B., ' Minimizing Structural Vibrations with Absorbers', Earthquake Engineering and Structural Dynamics, Vol.8, 219-236, 1980.

2. Chang, C. H. and Soong, Tsu T., ' Structural Control Using Active Tuned Mass Dampers', Journal of the Engineering Mechanics Division, ASCE, Vol. 106, No. EM6,

Dec. 1980.

3. J. P. Den Hartog, ' Mechanical Vibrations', 4th Ed., McGraw-Hill, New York, 1956.

4. Villaverde, R., ' Reduction in Seismic Response with Heavily-Damped Vibration Absorbers', Earthquake Engineering and Structural Dynamics, Vol.13, 33-42, 1985.

5. Warburton, G. B. and Ayorinde, E. O., ' Optimum Absorber Parameters for Simple Systems', Earthquake Engineering and Structural Dynamics, Vol.8, 197-217, 1980.

6. Warburton, G. B., ' Optimum Absorber Parameters for Minimizing Vibration Response', Earthquake Engineering and Structural Dynamics, Vol.9, 251-262, 1981.

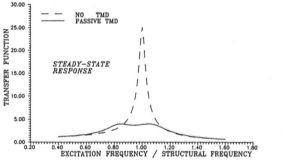

Fig 1

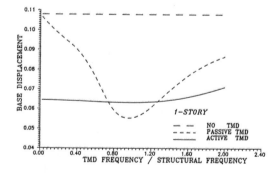

Fig 2

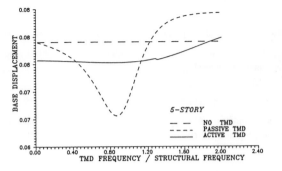

Fig 3

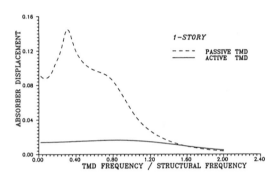

Fig 4

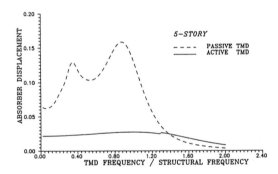

Fig 5

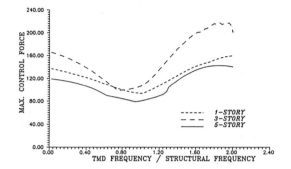

Fig. 6

A Neural Network Scheme for Adaptive Control of Large Structures

Daniel R. Rehak and Irving J. Oppenheim
Department of Civil Engineeering
Carnegie-Mellon University

Abstract

Large flexible structures will require control to eliminate vibrations. Consider a structure characterized by system output vectors and control force vectors that are arbitrarily long, where the system itself may be time-varying as the structure is modified or as the mass distribution changes under movements of people and equipment. *Adaptive control* is an appealing concept for such a problem because it accomplishes system self-identification and updating with system changes. The problem therefore is to apply adaptive control to an arbitrarily large structure as described. However, most adaptive control approaches remain model-referenced, whereas we wish to consider arbitrary structures for which such reference may be impossible; for instance, a structure in which the signal on any channel may be changed, lost, or reversed in sign. These are pathologies which defeat conventional control approaches, typically by making the system self-exciting.

A neural network (NN), acting as an I/O device with learning, may act as an adaptive controller for a system. An NN can also create stable representations for high-dimension vectors, thereby suggesting possible success in representing the space of a multiple-input multiple-output system. NN systems offer considerable robustness to corruption, and may scale up to arbitrarily large (massively parallel) systems. These understandings suggest that a NN approach may be found which would constitute a powerful architecture for the adaptive control of large systems. The research effort attempts the development of NN computation for that purpose, restricted to long period systems, for which learning can occur within time to correct conditions of self-excitation. This work is sponsored by the National Science Foundation under a *Small Grant for Exploratory Research* with a starting date of August 1, 1990.

Introduction and Problem Statement

In large flexible structures control (regulation) is required to eliminate vibrations induced by actions within and on the structure. A future vision, anticipating new materials and microelectronic technologies, suggests new properties which will lead to the broader use of such structures. As a hypothetical example of such properties, new materials may permit the costless monitoring of strain at closely spaced points in a large solid, or integrated sensors may provide monitoring of position (or acceleration) at every node in a truss. Similarly, new material properties may provide an inherent capability to actuate every strut in a truss or any discretized field of body forces in a solid. The two cases correspond to an arbitrarily long vector of system outputs and an arbitrarily long vector of control forces, respectively. It is recognized that all significant state variables can be observed and controlled, and therefore direct linear feedback should suffice as a control approach. However, new understandings in controls engineering are required if such structures are to be developed.

The system order is typically so large that modelling of the plant in state space is a challenging problem. A greater challenge is that the plant itself is time-varying as the structure is extended (built) or modified, or as the mass distribution changes with movements of people and equipment. Moreover, in a system with a great number of (inexpensive, redundant) sensors and actuators, individual "constants" will change and components will fail over time, but the control system should be robust with respect to such corruptions. *Adaptive control* is an appealing concept for such problems because it accomplishes system self-identification and updating with system changes. The new problem is to apply adaptive control to an arbitrarily large structure as described, and it is proposed that neural network computation be investigated for that purpose.

The problem statement in this investigation is first restricted to the vibration control of large (long period) structures and is next restricted to linearizable systems; the word is used in the control sense, referring to approximate linearity about an operating point. Such a scope embraces most anticipated large structures, still includes realistic material non-linearities, and still includes major influences such as discrete time-varying changes in the plant. At this stage of the work it excludes the non-linear problem of a grossly reconfigurable structure, and the practical application of vibration control under external dynamic forcing. We have made the initial restrictions to permit careful theoretical study of hypotheses; we believe that neural network performance is also well suited to forced systems and to the non-linear domain, but we prefer to restrict the scope of this first study.

Background

Neural network computation has a relatively long history, with fundamental contributions found in journals of biology, psychology, physics, and computer science. The field of neural network computation is expanding dramatically, as evidenced by the fact that three journals most closely related to neural network computation and possible engineering applications are all less than three years old: *Neural Networks*, which started publication in 1988, *Neural Computing*, 1989, and *IEEE Transactions on Neural Networks*, 1990. The reader can obtain an excellent current introduction to the field in two recent issues (September and October, 1990) of the *IEEE Proceedings*, and will find current research activities reported at the *International Joint Conference on Neural Networks* which is held once (or twice) yearly .

Recent neural network studies specific to the problem of control include one report [2] from Carnegie-Mellon (by Doo, Thewalt, and Rehak) on system identification and pulse control, containing simulation studies on one-, two-, and three-DOF elastic systems. A number of different network models are shown to perform well as system identifiers. Pulse control is then introduced, based first [6, 7] upon Udwadia's on-line, pulsed, open-loop adaptive control method to suppress vibrations for SDOF systems subjected to dynamic loading. For example, one neural network is used for system identification, and a second neural network is then used to learn pulse train trigger conditions. Simulation results with a variety of earthquake excitations show effective performance of the neural network as a pulse controller; Figure 1, from reference [2], presents the uncontrolled and the pulse-controlled displacement record under forcing by the Taft earthquake. A new simple pulse control scheme is then proposed, under the assumption that control forces can be applied in direct opposition to large inertial forces developed during seismic excitation, and simulation studies show it to be readily applicable and effective for MDOF systems; Figure 2 shows resulting displacement records and the ordered pulse trains, under the El Centro earthquake.

Much of the research on neural network capabilities for control has been motivated by their usefulness toward highly non-linear problems. Widrow has demonstrated many such control capabilities of neural networks, and in one recent paper [5] he demonstrates a "self-learning" system to learn required control intent for backing up a tractor-trailer. A recent paper [4] provides a general mathematical evaluation of neural computing for system identification and control, another [1] demonstrates identification capabilities, and another [3] discusses robotics applications including control.

Proposed Work

As stated above, neural networks, performing as I/O devices with learning, have been demonstrated to perform as adaptive controllers on low-order systems. Another NN finding is that a network can produce a stable representation for high-dimension vectors; this suggests that they may be suitable to represent the high-order space of large systems. Moreover, NN systems have demonstrated robustness to loss of connections or other corruptions. Finally, other NN teachings imply that important capabilities may scale-up readily to arbitrarily large (massively parallel) systems. Taken as a whole, these understandings suggest that a NN approach may be found which would constitute a powerful architecture for the adaptive control of large systems. However, no arguments or proofs have been offered and no conclusions can be drawn at present. Indeed, it is obvious that certain implementations (of a NN, a plant, and a controller) could be self-exciting and thereby unstable. The investigators seek fundamental analogues to prove or disprove NN capabilities for the problem as stated. If convincing physical arguments are found they will dictate an architecture, and they will then permit the analytical study of performance and bounds of scale. Such findings would be followed by simulation studies for comparison with predicted behavior. If all results are positive the work funded by this award will conclude with a design description for an experimental system to test the findings.

It is premature to fix and limit our theoretical hypotheses, but several can be noted briefly. It can first be reasoned that avoidance of self-excitation may be provided (in a broad sense) by satisfaction of biological feasibility. Turning next to conditions more directly associated with structural mechanics, one hypothesis to our work is that a NN, as an adaptive system, is shaped by two major factors: a convergence to an internalized energy-minimization which is most typically the end product of asynchronous excitation under learning, and an external coupling of input and output which is most typically a feedback condition in itself. A second hypothesis is that a NN for adaptive control requires a specific, and as yet unknown, structure for those external couplings; success in finding the full physical analogue will yield that required structure. A third theoretical observation is that the regulation of a structure results in minimization of an energy function of quadratic form, which leads us to hypothesize a NN model (again, exemplified physically by statistical mechanics) as close in form to the required adaptive controller. We believe such an outcome to be possible, and we believe that the resulting system performance will match well to the problem of regulating a large structure as described.

References

[1]   Chu, S.R., Shoureshi, R., and Tenorio, M.
      Neural Networks for System Identification.
      In *IEEE Control Systems Magazine*, pages 31-34. IEEE, April, 1990.

[2]   Doo, L.P., Thewalt, C., and Rehak, D. .
      Neural Networks for System Identification and Pulse Control of Linear Systems.
      Technical Report, Carnegie Mellon University, Department of Civil Engineering, Pittsburgh, PA,
           December, 1989.

[3]   Kung, S.-Y., and Hwang, J.-N.
      Neural Network Architectures for Robotic Applications.
      IEEE Transactions on Robotics and Automation 5(5):641-657, 1989.

[4]   Narendra, K.S., and Parthasarathy, K.
      Identification and Control of Dynamical Systems using Neural Networks.
      IEEE Transaction on Neural Networks 1(1), 1990.

p 198

[5]   Nguyen, D.H., and Widrow, B.
        Neural Networks for Self-Learning Control Systems.
        In *IEEE Control Systems Magazine*, pages 18-23. IEEE, April, 1990.

[6]   Udwadia, F. E., and Tabaie, S.
        Pulse Control of Single-Degree-of-Freedom System.
        Journal of Engineering Mechanics Division, ASCE 107(EM6):997-1009, 1981.

[7]   Udwadia, F. E., and Tabaie, S.
        Pulse Control of Structural and Mechanical Systems.
        Journal of Engineering Mechanics Division, ASCE 107(EM6):1011-1028, 1981.

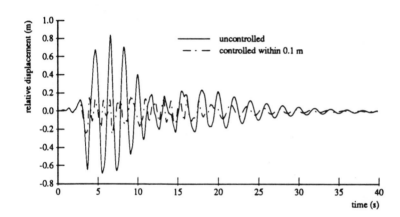

Figure 1. Response to Taft S69E Record (Taken from Ref. [2])

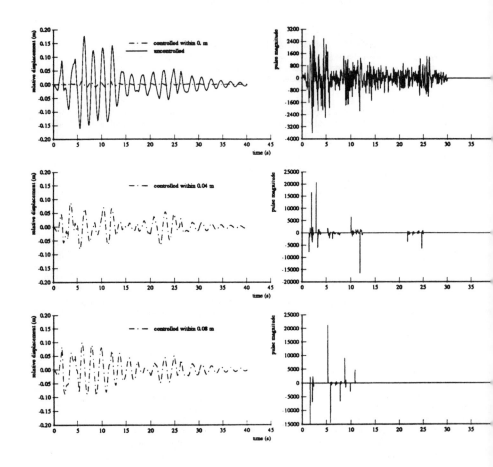

Figure 2. Response (Left) and Pulse Train (Right) to El Centro NS Record (Taken from Ref. [2])

FULL SCALE IMPLEMENTATION OF ACTIVE
BRACING FOR SEISMIC CONTROL OF STRUCTURES

A.M. Reinhorn and T. T. Soong
Department of Civil Engineering
State University of New York at Buffalo

INTRODUCTION

Active control using structural braces and tendons has been one of the most studied control mechanisms. Systems of this type generally consist of a set of prestressed tendons or braces connected to a structure whose tensions are controlled by electrohydraulic servo-mechanisms. One of the reasons for favoring such a control mechanism has to do with the fact that tendons and braces are already existing members of many structures. Thus, active bracing control can make use of existing structural members and thus minimize extensive additions or modifications of an as-built structure. This is attractive, for example, in the case of retrofitting or strengthening an existing structure.

A comprehensive experimental study was designed and carried out in order to study the feasibility of active bracing control using carefully calibrated structural models. In the first two stages, a 1:4 scale three-story model structure (3m tons, 2.5m high), modeling a shear frame building, was controlled using diagonal prestressed tendons activated by a servo-controlled hydraulic actuator. These experiments permitted a realistic comparison between analytical simulations and experimental results which were easily extrapolated to full-scale prototypes. Furthermore, important practical considerations of true delay, robustness of control algorithms, modeling errors and structure control system interaction could be identified and evaluated. In the third stage, a substantially larger and heavier six-story model structure (20m tons, 7.5m high) was controlled experimentally using multiple sets of diagonal active tendons producing substantial reduction of responses under various simulated earthquakes.

This paper focuses on the full scale implementation of active bracing system in an experimental six-story steel frame structure (600mtons, 15m high) constructed in a seismic prone area in Japan. The system was designed using extrapolated results from previous experimental studies and analytical simulations. The control system was assembled in the experimental structure, calibrated and tested by microtremors, by free vibrations of the structure and by forced vibrations supplied from the movement of an active mass damper used as an active excitater. The paper presents the highlights of the design and some initial test results.

EXPERIMENTAL SYSTEM - STRUCTURE AND ACTIVE BRACES

Test Structure

A dedicated full-scale structure was erected for performance verification of two control systems, i.e., active mass damper and active bracing system. Located in the city of Tokyo, Japan, the structure is a square, two bay six-story building, as shown in Figure 1. It was constructed using steel box tubular columns and W-shape beams with welded joints and bolted connections. The structure has a strong direction (1.02 sec period) and a weaker direction (1.5 sec period) to simulate a typical high rise building. The spectral properties of the structure obtained analytically are within 3% of the ones determined by field identification and are presented in Table 1. The damping of the structure was assumed to be 1% for analysis while the measured damping is actually 1% in the strong direction and only .5% in the weak direction.

Active Bracing System

A biaxial bracing system was assembled in the middle frames in the two orthogonal directions (see Figure 1). The bracing system consisted of two diagonal circular tubular braces attached within the first floor in each bay of the frame. A hydraulic actuator 312 kN capacity is assembled in each brace at the connection to the center column of the structure.

The actuators can expand and contract at a command from a digital controller (a microcomputer equipped with an INTEL 80386 processor) thus applying compression or tension in the braces and therefore introducing lateral forces into the structure. The computer command is a result of an on-line (real time) processing of measured data from velocity transducers placed in both directions at several floors in the structure. The hydraulic power for the actuators is supplied from a continuously operating small flow pump and a series of line accumulators which simply the large flow demands.

CONTROL ALGORITHMS

The real-time computations in the microcomputer are based on a reduced order model (ROM) active control algorithm. Two basic algorithms were designed for this investigation: (a) a classical closed loop linear optimal control low (Reinhorn, Soong et al, (1989)) modified to include only *velocities* measured at three critical floors in the structure (Soong, Reinhorn et al (1990)) and (b) a velocity feedback with observer which accounts for full dimensional state feedback with aid of a state estimator (Soong, Reinhorn et al (1990)).

Both algorithms were modified to include the compensation of delays in the force application in respect to the measured signal. The time delays are a result of phase shifts in instruments conditioners, time required for on-line computation and time required for the actuators to respond. A phase type compensation was employed in connection with a compensation of parallel time delays in computations and actuators. All compensations were based on field measured performances of control components (Reinhorn, Soong et al 1989).

DESIGN OF ACTIVE BRACING SYSTEM

Hydraulic System

The design of the active bracing system at the early stage was based on the experimental data obtained during the Stage 3 of the experimental study as described above. Due to the fact that similarity between the full-scale structure and the 1:4 scaled model is preserved, the sizing of the hydraulic actuator can be extrapolated directly from the model study. The primary parameters upon which the detail system design is based are the control force, the actuator displacement and the actuator velocity, which are related to the determination of the actuator capacity, the cylinder stroke, and the flow rate requirement of the hydraulic servovalve. In addition, the total flow of the hydraulic fluid required during a seismic event is the basis of sizing the hydraulic supply system. During the operating period of the active system, the hydraulic pomp supplies a constant flow of oil regardless of the actual requirement. When the demand of oil is more than what is supplied, the accumulator will provide a subsidiary flow to the actuator cylinder; inversely, when supply is more than demand, excess oil will be discharged into the accumulator, thus no additional pumping is necessary during the short period of earthquake excitation. For economic reasons, the pumping rate of the controller is determined to be the average flow rate estimated over the time history of the cumulative flow such that the accumulator volume is minimized. These design parameters were established in accordance with the simulated results of resource demands required for a desired structural performance under the design earthquake as discussed below.

Design Earthquakes

The local seismic records of the past seven years show velocities of less than 9.5 cm/sec. Accordingly, several earthquake accelerograms (El Centro N-S 1940, Hachinohe 1977, etc.) where scaled to a peak acceleration of 0.1g which corresponds to a criterion of 10cu/sec maximum velocity. Response analyses were performed and required control force and force rates were carefully recorded to verify the adequacy of design specification.

Hydraulic Actuators Design

The maximum required control force obtained in the simulation studies was used to size the hydraulic actuators. The maximum required diagonal force approaches 9% of the total weight of the structure and produces a reduction of 40% of the earthquake response which is equivalent with an 7% increase in the critical damping.

It should be noted that the maximum force is required for several fractions of seconds which do not impose a tremendous demand on hydraulic power.

Hydraulic Accumulators

Since the total calculated oil volume flowing during an entire earthquake episode does not exceed two gallons (approximately 7570 cm^3), the total flow can be supplied from a hydraulic accumaulator and a small volume pump. The hydraulic accumulators were sized to supply the required volume of oil at high flow peak demands while the pump works to supply the losses throughout the servovalves and hydraulic pipes along the small percentage of the toal required volume of oil (see Fig.2). A final selection of 10 gallons (37850 cm^3) accumulators (volume includes also the pressurized nitrogen which maintains the line pressure) was made.

Simulated Results

The simulations using the design earthquakes produced satisfactory results as shown in Table 2. The total power required during an earthquake reaches 20 kw for most earthquakes except for Hachinohe earthquake where 50 kw are required. The total energy consumption during an earthquake of one minute will be of the order of 0.3 kwh to 0.9 kwh which can be delivered from passive electrical power sources. Additional power should be considered for the continuous operation of charging electrical and hydraulic equipment.

PERFORMANCE OF CONTROL SYSTEM

Expected Structural Response

A typical structural response of the full-scale system is shown in Figure 3.. The performance is shown for the top floor of the structure where both displacements and accelerations are reduced. As a result, the base shear is also substantially reduced. The required resources to produce such performance are shown in Figure 4 (for the strong direction of the structure for both actuators operating in this direction). The large forces are required for short period of times associated with several power peaks and a rapid initial energy demand.

Experimental Performance of Control System

Recently the construction was completed and initial tests were performed on the site. The tests included three stages: (a) control of microtremor excitation due to topic and wind; (b) control of free vibrations of the structure after excitation produced by the uses of the active mass damper as a force exciter; and (c) control of a forced vibration using the active mass damper exciter.

The performances of the structure in free vibration is shown in Figure 5. This vibration indicates extremely low damping in one direction and low damping in the other. Figure 6 shows the build-up of the structural vibration and then the influence of control applied at the peak of excitation. The increased damping is evident (from 0.5% to 3.5% in the weak direction). Figure 7 shows the influence of control during forced vibrations. The response builds up in the transient stage and reaches its steady state while is evidently smaller than the previous tendency as shown in Figure 6.

CONCLUSIONS

The paper presents a full-scale implementation of an active bracing system which was constructed and tested recently. The analytical simulations and the structural testing indicate that the system can produce reduction of seismic response which might be difficult to obtain by other measure. More issues, however, need to be addressed: (a) a more efficient design of power supply; (b) a simplification of actual components for practical applications; (c) fail safe and back-up systems; (d) maintenance issues and; (e) reliability issues.

ACKNOWLEDGEMENTS

This work was performed using grants from the National Center for Earthquake Engineering Research which in turn is supported by NSF and NYSERDA. The authors acknowledge the support from Takenaka Construction Company of Tokyo, Japan, Kayaba Industries, Tokyo, Japan and MTS Systems Corporation of Minneapolis, MN. The authors wish to acknowledge the contribution of Dr. R.C. Lin, Mr. P. Wang, Mr. M. Riley, Mr. M. Pitman and Mr. D. Walch who participated in various stages of this project.

REFERENCES

Reinhorn, A.M. and Soong T.T., et al. (1989), "1:4 Scale Model Studies of Active Tendon Systems and Active Mass Dampers for Aseismic Protection," Report NCEER-89-0026, National Center for Earthquake Engineering Research, Buffalo, NY.

Soong, T.T. (1990), *Active Structural Control: Theory and Practice,* Longman, London and Wiley, NY.

Soong, T.T., Reinhorn, A.M., Wang, Y.P. and Lin, R.C. (1990), Full-Scale Implementation of Active Structural Control Under Seismic Loads I. System Design and Simulation Results, ASCE/Journal of Engineering Mechanics, (submitted).

	X-direction		Y-direction	
Mode	Record (Hz)	Analysis (Hz)	Record (Hz)	Analysis (Hz)
1	0.900	0.922	0.650	0.643
2	2.700	2.703	1.800	1.750
3	4.650	4.737	2.975	2.918
4	6.850	6.973	4.125	4.117
5	9.400	9.481	5.425	5.511
6	12.75	13.03	7.250	7.623

Table 1. Dynamic Characteristics of Test Structure

Earthquakes	Scale factor		Performance indices			Resource requirement / Actuator						
			Top fl. disp. (cm)	Top fl. acc. (g)	Base shear (kN)	control Force (kN)	Actuator disp. (cm)	Max. flow rate (gpm)	Ave. flow rate (gpm)	Vol.of accum. (gal.)	Total volume of oil (gal.)	Power (Kw)
El Centro	0.32	Unctrl	7.98	0.37	1018.3	313.9	0.51	19.72	2.7	0.11	0.74 in 20 s	20.02
		Ctrl	4.66	0.23	618.0							
		Red.(%)	41.6	36.5	39.5							
Miyagioki	0.68	Unctrl	5.30	0.43	847.6	245.3	0.40	15.24	2.0	0.12	0.71 in 30 s	16.08
		Ctrl	3.27	0.21	547.4							
		Red.(%)	36.4	50.7	35.8							
Mexico (N90W)	0.65	Unctrl	9.70	0.32	1347.9	255.1	0.41	6.21	1.2	0.18	1.33 in 100 s	5.65
		Ctrl	5.80	0.18	877.0							
		Red.(%)	40.2	44.2	35.0							
Mexico (S00E)	1.115	Unctrl	9.95	0.34	1389.1	323.7	0.53	9.36	1.5	0.34	2.13 in 100 s	10.77
		Ctrl	6.99	0.24	1000.6							
		Red.(%)	29.8	30.4	28.2							
Pacomia Dam (S16E)	0.095	Unctrl	5.58	0.35	741.6	255.1	0.42	10.94	2.0	0.06	0.47 in 20 s	10.06
		Ctrl	3.90	0.17	500.3							
		Red.(%)	30.1	50.7	32.7							
Pacomia Dam (S74W)	0.104	Unctrl	4.14	0.30	606.3	176.6	0.29	9.35	2.0	0.055	0.44 in 20 s	4.88
		Ctrl	2.68	0.14	406.1							
		Red.(%)	35.1	52.6	33.0							
Taft (N21E)	0.715	Unctrl	7.64	0.63	1083.0	333.5	0.54	18.29	3.0	0.12	0.96 in 20 s	21.48
		Ctrl	4.36	0.30	535.6							
		Red.(%)	42.9	53.3	50.5							
Tokyo	1.48	Unctrl	4.85	0.33	623.9	333.5	0.54	12.63	2.7	0.07	0.37 in 10 s	21.60
		Ctrl	3.73	0.26	406.1							
		Red.(%)	23.2	21.1	34.6							
Hachinohe	0.60	Unctrl	17.93	0.79	2731.1	696.5	1.13	26.74	6.0	0.17	1.51 in 20 s	59.02
		Ctrl	9.48	0.41	1471.5							
		Red.(%)	47.1	48.6	46.0							
Hachinohe (limited Ctrl force)	0.60	Unctrl	17.93	0.79	2731.1	333.5	0.54	31.04	4.0	0.17	1.21 in 20 s	55.54
		Ctrl	12.65	0.69	1660.8							
		Red.(%)	29.4	12.5	39.2							

Table 2. Performance and Resources for Control of Seismic Motions

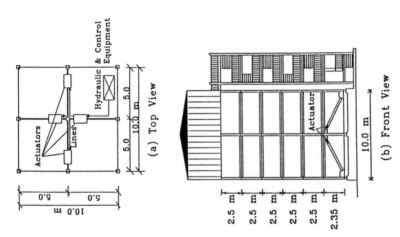

Fig. 2. Flow Balance in Hydraulic System

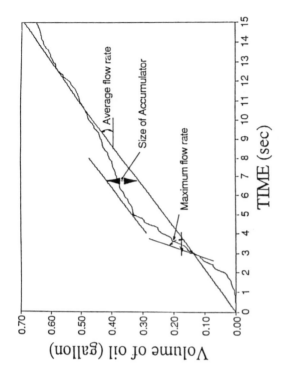

Fig. 1. Elevations of Full
Scale Structure

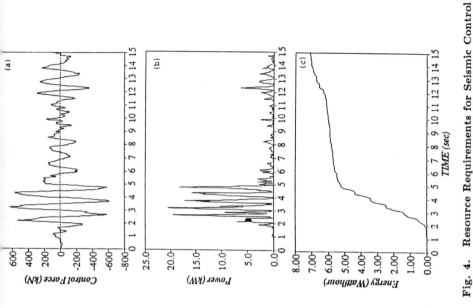

Fig. 4. Resource Requirements for Seismic Control

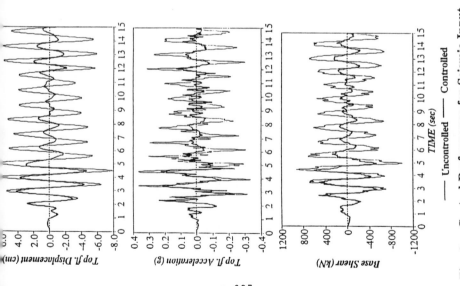

—— Uncontrolled ——— Controlled

Fig. 3. Control Performance for Seismic Input
(El-Centro 1940)

p 207

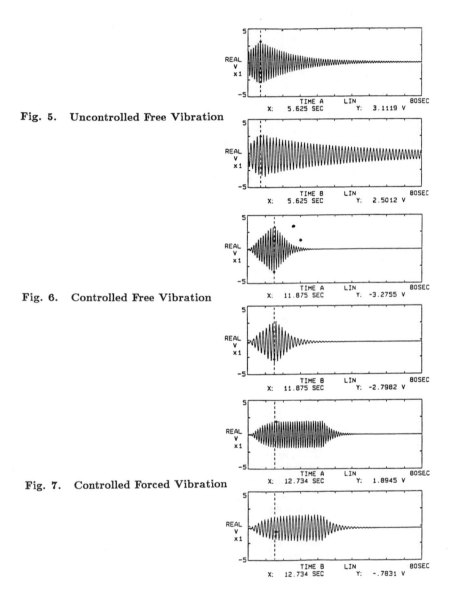

Fig. 5.   Uncontrolled Free Vibration

Fig. 6.   Controlled Free Vibration

Fig. 7.   Controlled Forced Vibration

p 208

Frequency Domain Optimal Control Design for Civil Engineering Structures

Extended Summary for the U.S. National Workshop on Structural Control Research

J. Suhardjo* and B. F. Spencer, Jr.[†]

Introduction

The concept of structural control was first presented to the structural engineering profession by Yao [7]. In this work, Yao implied that performance and serviceability of a structure could be controlled so that they remain within prescribed limits during the application of environmental loads. To ensure safety, the displacements of a structure should be limited, whereas for the comfort of people who use the structure, accelerations need to be limited. Recognizing the potential benefits of active structural control for civil engineering applications, researchers have conducted extensive studies in the last two decades. Recently, T. T. Soong [3] reviewed the state of the art in active structural control for civil engineering applications and cited over 120 articles. Control strategies have been reported for truss type structures, suspension bridges, tall buildings, offshore structures, *etc.* In addition, important experimental studies have been conducted to demonstrate the feasibility of active control systems to mitigate dynamic hazards in civil engineering structures [4,1,2].

Control design for civil engineering structures to date has been primarily performed in the time domain. However, identification and modeling of the dynamic behavior of the structure, as well as associated excitations such as winds, waves, and earthquakes, are usually done in the frequency domain. As a result, studies have been initiated to explore the possibility of using frequency domain based optimal control techniques for the design of controllers in civil engineering structures [5,6].

In this paper, recently developed frequency domain optimal control techniques will be presented. An example will be included to illustrate features of these methods and their utility for structural control design in the presence of uncertainty will be discussed.

Problem Formulation

Consider the idealized n-degree-of-freedom structure subjected to an external load, for example earthquake ground excitation. This system can be represented in the state-space representation as

$$\dot{x} = Ax + Bu + N\ddot{x}_g ; \quad y = Cx$$

where x is a state vector of displacements and velocities, u is a control force vector, y is a measured output vector, $\ddot{x}_g$ is the ground acceleration, A is a matrix of structural parameters, B is a matrix specifying the location of control forces, N is the matrix specifying how

*Post-Doctoral Fellow, Dept. of Civil Engrg., Univ. of Notre Dame, Notre Dame, Indiana 46556.
[†]Associate Professor, Dept. of Civil Engrg., Univ. of Notre Dame, Notre Dame, Indiana 46556.

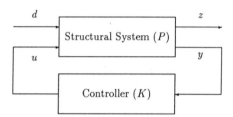

d z

Structural System (P)

u y

Controller (K)

Figure 1: Basic structural control system.

the ground acceleration enters the structure, and C is a vector specifying the location of measurement sensors.

Using a block diagram problem description, the structural control problem can be depicted as in Figure 1, where y is the measured output vector, z is the regulated output vector, u is the control input vector, and d is the exogenous input vector or disturbance excitation vector. The regulated output vector z can consist of any combination of states of the system, including accelerations, as well as components of the control input vector u.

Appropriate weighting functions can also be added to elements of z to designate the frequency range over which each element of z is to be minimized and can be tailored to suit the physical properties of the system. When the regulated output is the structural response, a weighting function specifying desired performance should be large in the frequency range of the lowest natural frequencies and small at high frequencies. On the other hand, when the regulated output is the control force, the weighting function should be small at low frequencies and large at high frequencies. Increasing one of the weightings in a frequency range decreases the importance of the other weighting in that frequency range.

The task here is to design a stabilizing controller in such a way that a norm or "size" of the transfer function from d to z is minimized. Control design using the H_2 and H_∞ norms to measure the "size" of transfer functions has attracted much recent research attention. As suggested by the name, H_2 control design is a control design which searches for a stabilizing controller while minimizing the H_2 norm of the transfer function from d to z, $\|H_{zd}\|_2$, while H_∞ control design minimizes $\|H_{zd}\|_\infty$. The next section will provide an example of these control design techniques.

Example

To illustrate the use of a block diagram problem description, consider the problem of controlling a structure under earthquake excitation. One possible block diagram set-up for P is depicted in Figure 2. In this figure, A, B, and C are matrices from the state-space representation the equations of motion for the structure, and the block F represents the earthquake model. The output of F is g, a vector which consists of zeros and ground accelerations, $\ddot{x}_g$. Note that in general the control force u and the ground acceleration $\ddot{x}_g$ enter the system differently. For example, for a multistory building the way a control force comes into the system is dictated by the placement of active control devices while ground acceleration $\ddot{x}_g$

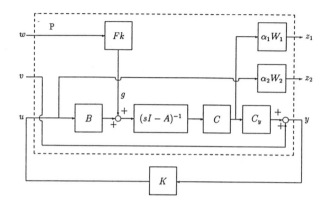

Figure 2: A control problem for structure under earthquake excitation.

comes into every floor of the building. The external excitation d consists of two elements: w and v, or $d = (\ w \quad v\ )^T$, where w is a white noise input to the earthquake model filter and v is the measurement noise. The regulated output z consists of z_1 and z_2 (*i.e.* $z = (\ z_1 \quad z_2\ )^T$) where z_1 is the weighted structural response and z_2 is the weighted control force. W_1 and W_2 are frequency dependent functions, α_1, α_2 are scalar multipliers for W_1, W_2, respectively. The scalar k expresses a preference in minimizing the transfer function from w to z versus minimizing the transfer function from v to z. C_y is a matrix which determines which combinations of state variables are measured.

Let us now consider a SDOF system with mass $m = 10$ kg, natural frequency $\omega_0 = 20$ rad/s, and damping coefficient $\zeta_0 = 0.01$. The earthquake model F is a white noise filter. To illustrate the flexibility of this frequency domain approach, let $Cy = (\ -\omega_0^2 \quad -2\zeta_0\omega_0\ )$ so that the measured output is the acceleration. The weighting functions W_1 and W_2 are chosen in such a way that the regulated output of the system are the displacement and control force. The basic weighting function W_1 for this example is depicted in Figure and W_2 is chosen to be unity. With these W_1 and W_2, the regulated output z_1 is the weighted displacement and z_2 is the control force.

The statistics of the structural responses and control force are calculated from stochastic differential equations of the system. For the H_2 control, the standard deviation of the displacement is 3.6887×10^{-2} m and the standard deviation of the control force is 5.1187×10^1 N. For the H_∞ control, the standard deviation of the displacement is 2.4210×10^{-2} m and the standard deviation of the control force is 7.1024×10^1 N. For comparison, the standard deviation of displacement of uncontrolled system is 1.4012×10^{-1} m.

Figure 4 shows the transfer functions from the white noise excitation w to the displacement response for both the H_2 and H_∞ controlled systems. Another interesting thing is that there is a tradeoff between the H_2 and H_∞ control designs. The advantage offered by the H_∞ control in minimizing the peak of a transfer function has to be paid by disadvantages in having higher value of the transfer function at other frequencies.

p 211

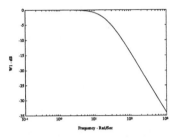

Figure 3: Basic weighting function W_1

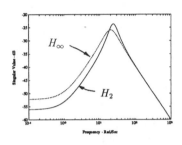

Figure 4: Comparison of H_2 and H_∞ Transfer Functions.

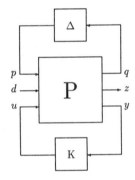

Figure 5: The block diagram representation of a system with uncertainty model.

Control Design Incorporating Uncertainty

The H_2 and H_∞ control methods described above can be used to design robust controllers. In this case, a model of uncertainty is incorporated in the problem formulation, as depicted in Figure 5. In this figure, Δ is the uncertainty model, p and q are points where the uncertainty enters and leaves the system.

There are two types of robustness of a control system: stability robustness and performance robustness. Stability robustness is the ability of a control system to remain stable in the face of model uncertainty. Performance robustness is the ability of a control system to perform within prescribed performance limits in the face of model uncertainty. Stability robustness is more straight-forward because there is a clear criterion: the control system is either stable or not stable. Performance robustness is a more difficult subject and requires quantitative measure of performance.

A preliminary study of the design of a robust structural control design for stability has been conducted by Suhardjo [5]. In that work, it is shown that to improve the stability robustness of a controlled system, the norm of the transfer function from p to q, H_{qp}, has to be minimized. In order to minimize the norms of H_{zd} and H_{qp} at the same time, p is appended to d and q is appended to z to obtain d_Δ and q_Δ where

$$d_\Delta = \begin{pmatrix} d \\ p \end{pmatrix} \; ; \;\; z_\Delta = \begin{pmatrix} z \\ q \end{pmatrix}.$$

The stability robust control problem is then shown to be equivalent to finding a stabilizing controller K to minimize the appropriate norm of $H_{z_\Delta d_\Delta}$. By doing this, trade-offs between the desired performance of the nominal system and the stability of the system with uncertainty can be considered.

Conclusions

In this paper, recently developed frequency domain strategies such as the H_2 and H_∞ synthesis have been presented. These techniques provide significant physical insight into structural control problems. The freedom of the control designer to choose elements of regulated output and frequency dependent weighting functions makes these techniques very flexible and opens up many research possibilities. An example is given utilizing the H_2 and H_∞ control designs in which accelerations measurements were employed. Application of these techniques to robust control has also been discussed.

References

1. Chung, L.L., Reinhorn, A.M., and Soong, T.T., "Experiments on Active Control of Seismic Structures," *Journal of Engineering Mechanics,* ASCE, Vol. 114, No. 2, 1988, pp. 241–256.
2. Reinhorn, A.M. *et al.* , "1:4 Scale Model Studies of Active Tendon Systems and Active Mass Dampers for Aseismic Protection," Report No. NCEER 89–0026, *National Center for Earthquake Engineering Research,* Buffalo, New York, 1989.
3. Soong, T.T., "State-of-the-Art Review: Active Control in Civil Engineering," *Engineering Structures,* Vol. 10, 1988, pp. 74–84.
4. Soong, T.T. and Skinner, G.T., "Experimental Study of Active Structural Control" *Journal of Engineering Mechanics Division,* ASCE, Vol. 113, No. EM6, 1981, pp. 1057–1067.
5. Suhardjo, J., *Frequency Domain Techniques for Control of Civil Engineering Structures with Some Robustness Considerations,* Ph.D. Dissertation, University of Notre Dame, Department of Civil Engineering, 1990.
6. Suhardjo, J., Spencer, B.F., Jr. and Kareem, A., "Active Control of Wind Excited Buildings: A Frequency Domain Based Design Approach,", submitted to the *Eight International Conference on Wind Engineering,* July 8–12, 1991, London, Ontario.
7. Yao, J.P.T., "Concept of Structural Control," *Journal of the Structural Division,* ASCE, Vol. 98, No. ST7, 1972, pp. 1567–1574.

ACTIVE VIBRATION CONTROL FOR HIGH-RISE BUILDINGS USING DYNAMIC VIBRATION ABSORBER DRIVEN BY SERVO MOTOR

T. SUZUKI, M. KAGEYAMA, A. NOHATA, M. SEKI, A. TERAMURA and T. TAKEDA
Obayashi Corporation Technical Research Institute
4-640, Shimokiyoto, Kiyose-shi, Tokyo, 204, JAPAN

K. YOSHIDA and T. SHIMOGO
Keio University, 3-14-1, Hiyoshi, Kohoku-ku, Yokohama, 223, JAPAN

ABSTRACT

An active dynamic vibration absorber (DVA) driven by an AC servo motor through a ball screw is developed in this study. The facilities of this DVA are smaller than the usual ones controlled by a servo hydraulic actuator. The optimization of the controller is performed taking account of the dynamic characteristics of the active DVA driven by a speed control driver. By carrying out digital computer-controlled type tests and numerical calculations using a 4-story, 11-ton weight model frame structure, it is demonstrated that the active DVA, with the absorber's mass ratio to the structure only 0.5%, can reduce the response of every story, not only displacement but also acceleration, to 1/3 compared with no-control. As a result, the usefulness and feasibility of application of this active DVA in high-rise buildings have been made clear.

1. INTRODUCTION

An active vibration control system which actively controls vibrations in medium and small-scale earthquakes and under strong winds has been developed for the purpose of maintaining functional properties and improving residential conditions of high-rise buildings. Among features of this system are included, ① that from the standpoints of saving space occupied by the apparatus and its accessory equipment, and effectiveness of multi-mode control, an active dynamic vibration absorber which drives added mass supported by linear bearings with an AC servo motor through a ball screw is utilized, and ② as the control technique, upon giving consideration to the dynamic characteristics of the AC servo motor [1], [2], the probabilistic optimum control theory for a multi-degree-of-freedom vibration system has been employed.

This report gives an outline of the vibration control system and describes the results of vibration control experiments by the feedback system conducted using a model building.

2. EXPERIMENTATION MODEL

2.1 Specifications of Experimentation Model

The specifications of a 4-storied structure equipped with an active dynamic vibration absorber (DVA) at its top-most part and the DVA are shown in Fig.1. The total weight of the main structure is approximately 11 ton and the mass of the active DVA 50 kg (mass ratio≒ 0.5%). The base of the structure is attached rigidly to a shaking table, and because of this, rocking occurs influenced by the rigidity of the shaking table. When the degrees of freedom of rocking is included, the 4 -storied structure has five degrees of freedom and the natural frequencies from primary to quinary are 1.86 Hz, 6.19Hz, 10.0 Hz, 13.1 Hz, and 27.3 Hz, respectively. The damping factor of the primary mode is approximately 0.5 %.

2.2 Dynamic Characteristics of AC Servo Motor

A conceptional drawing of the composition of the active DVA used in these experiments is given in Fig.2. The AC servo motor used as the actuator has a rated output of 1.5 KW in view of balance of the dynamic vibration absorber mass and the control force required, and is driven by a driver which provides speed control by minor loop. In this study, the frequency response of velocity ν of the DVA mass in relation to input voltage u to the speed control driver of the AC servo motor was experimentally investigated, and this was approximated by the following quadratic equation :

$$\frac{V(s)}{U(s)} = \frac{-\beta}{S^2 + \alpha_1 S + \alpha_2} \qquad (1)$$

where,

$$V(s) = L[\nu(t)], \quad U(s) = L[u(t)]$$

provided that $L[\cdot]$: Laplace transform

Accordingly, the equation of motion of the active DVA including the speed control driver can be approximated as follows :

$$\ddot{x}_a(t) = -\alpha_1 \ddot{x}_a(t) - \alpha_2 \dot{x}_a(t) - \beta u(t) \qquad (2)$$

provided that x_a is relative displacement of active dynamic vibration absorber mass.

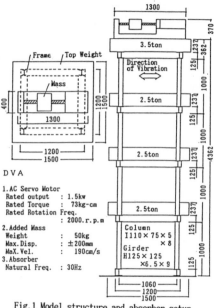

DVA

1. AC Servo Motor
Rated output : 1.5kw
Rated Torque : 73kg-cm
Rated Rotation Freq. : 2000.r.p.m
2. Added Mass
Weight : 50kg
Max.Disp. : ±200mm
MaX.Vel. : 190cm/s
3. Absorber
Natural Freq. : 30Hz

Colum I110×75×5 ×8
Girder H125×125 ×6.5×9

Fig.1 Model structure and absorber setup

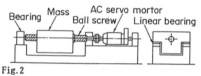

Fig.2 Schematic of active dynamic vibration absorber

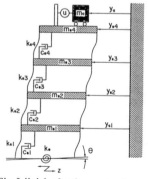

Fig.3 Model of primary structure

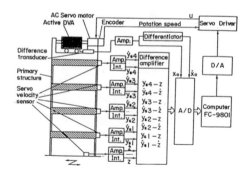

Fig.4 Schematic of control system

2.3 Formulation of Mechanical Model

The mechanical model of this experiment is shown in Fig.3. With relative displacement vector of the structure $x_s = y_s - e z$, and the relative displacement of the DVA as $x_a = y_a - \Delta^T y_s$, the state form equation of this total system including the characteristics of the DVA will be as in Eq. (3), where a newly defined state vector expression as in Eq. (4) is considered.

$$\dot{x}(t) = A\ x(t) + B\ u(t) + D\ \ddot{z}(t) \tag{3}$$

Where, $e^T = [\ 1\ 1\ 1\ 1\ 0\ ]$, $\Delta^T = [\ 1\ 0\ 0\ 0\ 0\ ]$

$$x^T(t) = [\ x_a(t),\ \dot{x}_a(t),\ \ddot{x}_a(t),\ x^T_s(t),\ \dot{x}^T_s(t)\ ] \tag{4}$$

where, $x^T_s(t) = [\ x_4,\ x_3,\ x_2,\ x_1,\ \theta\ ]$

3. OPTIMIZATION OF ACTIVE DVA

With the relative displacement and the absolute acceleration of the main structure separately made objective functions, and further, with the stroke of the DVA mass and input voltage to the AC servo motor as restrictive conditions, the quadratic form evaluation criterion may be expressed as the evaluation functions of the following two equations :

Evaluation Function I : Relative Displacement

$$J_d(u) = E\ [\ x^T Q_d\ x + r_d\ u^2\ ] \tag{5}$$

Evaluation Function II : Absolute Acceleration

$$J_a(u) = E\ [\ \ddot{y}^T_s\ Q_a\ \ddot{y}_s + r_a\ u^2\ ] \tag{6}$$

provided that, $E[\ \cdot\ ]$ expresses the mathematical expectation. Where x and $\ddot{y}$ are the state vector, u is the control vector, and Q_d, r_d, Q_a, and r_a express weight coefficient matrices.

The optimum control laws for minimizing the evaluation function equations (5) and (6), may be given as follows by feedback gain vectors F_d and F_a using the solutions of the respective algebraic Riccati equations :

Evaluation function I : $\quad u_d(t) = F_d\ x(t) \tag{7}$

Evaluation function II : $\quad u_a(t) = F_a\ x(t) \tag{8}$

For realization of the optimum control laws (7) and (8), it will be necessary for feedback of all state quantities, but in this study, it was considered that the state quantity of rocking and the quantity of acceleration of the DVA mass would not be fed back, and control was made reduced-order by recalculating feedback gain based on the quasi-optimization method [1] using output of feedback control proposed by Nishimura et al.

4. COMPOSITION OF VIBRATION CONTROL SYSTEM

The composition of the vibration control system is shown in Fig.4. This system is composed of various sensors, a dynamic vibration absorber, and a control panel governing the entire system, where a computer FC-9801 is accomodated inside. The relative displacements of the structure from the foundation are determined by integrating the velocity signals obtained from velocimeters installed at the individual stories and the foundation, and transforming them into displacement signals and taking the differences. Relative velocities of the structure are similarly determined from velocity signals. The relative displacement of the DVA mass is detected using a differential transducer, and its relative velocity is obtained passing this through a differential analyzer. These detected quantities are analog-to-digital transformed at sampling period T = 5 ms and sent to the computer, and after operation of the control law in the computer, they are impressed on the AC servo motor through a digital-to-analog transducer and speed control driver. The rotating force of the AC servo motor is transmitted to a ball screw to cause the added mass supported by linear bearings to be horizontally displaced to produce the control force required for the control system.

5. VIBRATION CONTROL EXPERIMENTS BASED ON EVALUATION FUNCTION I

5.1 Experimentation Method

As forced vibration experiments, sinusoidal excitation experiments to investigate the frequency amplification characteristics of response accelerations of the structure against input acceleration, and non-stationary random wave excitation experiments using El Centro NS seismic wave to investigate the responses to seismic wave input were carried out. The El Centro wave was applied in real time with the maximum value as approxiamtely 50 gal, and to eliminate the influences due to the dynamic characteristics of the shaking table and the structure, the waveform itself applied to the shaking table was compensated by digital signal treatment using high-speed Fourier transform.

In determination of weight coefficients Q_d and r_d, Q_d was made constant, and aiming for the damping factor of the foundamental vibration mode becoming around approximately 15 %, the weight coefficient r_d (r_d = 2E-0.7) concerning control force was selected. Hereafter, the no-control experiment and control experiment at r_d = 2E-0.7 will be referred to as CASE D.0 and CASE D.1, respectively. Further, aiming for the above mentioned mode damping factor to be about 2 to 3%, a control experiment with weight coefficient made r_d = 2E-0.6 and with control force made smaller will be referred to as CASE D.2.

5.2 Sinusoidal Input Experiment Results

The acceleration frequency response amplification factor and the phase difference of the third story for the input acceleration in the sinusoidal input experiments are shown in Fig.5.

Cases of control by active DVA (CASE D.1, r_d =2E-0.7) are indicated by ◯ marks, and cases of no-control (CASE D.0) by △ marks. It may be comprehended from Fig.5 that by exerting control, resonance has been suppressed well in the primary and secondary modes. With regard to the primary mode, whereas a response amplification of approximately 100 times had been indicated with no-control, this was held to approximately 4.0 times through control, while in the secondary mode, approximately 12 times was held to 1.0 time.

5.3 Seismic Wave Input Experiment Results

The results of non-stationary random wave excitation experiments using El Centro NS waves are shown in Table 1 and Fig. 6 and 7. Table 1 gives comparisons of maximum response values and RMS responses in CASES D.0, D.1, and D.2, while Figs.6 and 7 respectively show the response waveforms at the third-story location and responses of the DVA mass in CASE D.0 and CASE D.1.

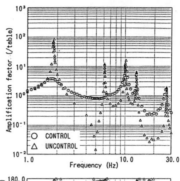

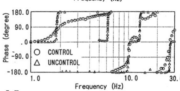

Fig.5 Frequency response functions of third-story acceleration to input acceleration (CASE D.0 and CASE D.1)

Table 1 Comparison of maximum and RMS responses (CASE D.0 , CASE D.1 and CASE D.2)

		CASE — D.0 Uncontrol		CASE—D.1 r =2E-07		CASE—D.2 r =2E-06	
		Max.	rms	Max.	rms	Max.	rms
Disp. (cm)	4	1.95	0.271	0.383	0.047	0.958	0.077
	3	1.55	0.222	0.308	0.035	0.765	0.049
	2	1.06	0.134	0.203	0.026	0.548	0.038
	1	0.522	0.073	0.110	0.018	0.283	0.021
Acc. (gal)	4	235.	41.6	127.	10.9	183.	16.40
	3	211.	32.9	63.5	7.04	108.	9.85
	2	199.	20.6	53.8	5.04	121.	10.46
	1	120.	14.5	57.1	5.42	64.4	9.18
Input Acc. (gal)		50.7	---	53.7	---	63.0	---

According to Table 1 and Figs. 6 and 7, in spite of the DVA's mass ratio being low at 0.5%, large vibration control effects have been obtained on both acceleration and displacement responses of the various stories. Further, since compensation of the input waveform was not sufficient, there is a slight difference in input between non-controlled and controlled, but when numerical corrections are made based on frequency response functions under identical input conditions, the maximum response value in CASE D.1 will be approximately 1/4 in terms of displacement and approximately 1/3 in terms of acceleration compared with the no-control experiment of CASE D.0.

5.4 Simulation Analysis Results

The results of simulation analyses of no-control (CASE D.0) and control (CASE D.1) experiment are given in Table 2. Simulation analysis results of the response waveforms for the third-story location and responses of the DVA in control experiment are shown together with experimental results in Fig. 6 and 7, respectively.

According to Table 2 and Fig. 6 and 7, the analytical and experimental results agree well with respect to both structure responses and DVA responses, and the appropriateness of the analysis technique is verified.

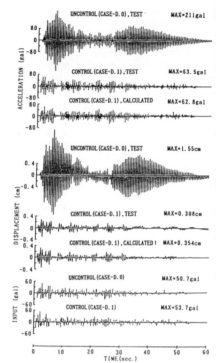

Fig.6 Test and calculated response waveformes
(CASE D.0 and CASE D.1)
(El Centro Earthquake input)

Table 2 Comparison between test and caluculated results (CASE D.0 and CASE D.1)

		CASE-D.0		CASE-D.1		
		Test	Analysis	Test	Analysis	
Disp. (cm)	4	1.95	1.75	0.383	0.407	Behavior of Dynamic Vibra- tion Absorber
	3	1.55	1.49	0.308	0.354	(Test)
	2	1.06	1.07	0.203	0.257	Disp.=13.1cm Acc.=2650gal
	1	0.522	0.570	0.110	0.139	(Analysis)
Acc. (gal)	4	235.	251.	127.	85.3	
	3	211.	219.	63.5	62.8	Disp.=13.5cm Acc.=2010gal
	2	199.	189.	53.8	50.5	
	1	120.	141.	57.1	44.0	
Input Acc. (gal)		50.7	50.7	53.7	53.7	

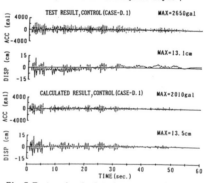

Fig.7 Test and calculated response of absorber
(CASE D.1)
(El Centro Earthquake input)

6. VIBRATION CONTROL EXPERIMENTS BASED ON EVALUATION FUNCTION II

6.1 Experimentation Method

As experiments, similarly to the case of Evaluation Function I, sinusoidal excitation experiments and non-stationary random wave excitation experiments using El Centro NS wave were conducted. In order to make the features of control by acceleration evaluation more distinct, the El Centro waves were applied shortening the time axis to 1/3, and the experiments were carried out making the high-order modes of the structure more easily excited.

In determining the weight coefficients Q_* and r_* , Q_* was made constant, and r_* was selected aiming for the damping factor of the fundamental vibration mode to be of about the same value as for the displacement evaluation experiment, CASE D.2, that is, 2 to 3 %. The reason the control force was set up to be comparatively small, was that it was not possible to make the control force large in the acceleration evaluation experiment because of a problem of accuracy of velocity sensors in the low frequency range.

Hereafter, the no-control experiment with input made reducing the time axis to 1/3 will be referred to as CASE A.0, the acceleration evaluation control experiments as CASE A.2, and the displacement evaluation experiment according to the weight coefficients described in Section 5.1 as CASE D.2.

6.2 Sinusoidal Input Experiments

The acceleration frequency response amplification factor and the phase difference of the third story for the input accleration in the sinusoidal input experiments are shown in Fig.8. Acceleration evaluation control (CASE A.2, Δ marks) and no-control (CASE A.0, ◯ marks) are compared in the figure. Compared with the case of displacement evaluation control (see Fig.5), resonance at high order, especially the tertiary mode was suppressed. Along with which, in acceleration evaluation control, the results were roughly the same as with no-control with regard to frequency ranges of low response amplification factors where the results exceeded those of no-control in displacement evaluation control, and the feature of the acceleration evaluation function appeared distinctly. The primary resonance amplification factor of the CASE A.2 experiments was approximately 25 times, corresponding roughly with the before-mentioned damping factor value set up.

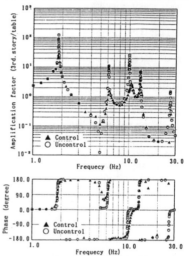

Fig.8 Frequency response functions of third-story acceleration to input acceleration (CASE A.0 and CASE A.2)

Table 3 Comparison of control methods (CASE A.2 and CASE D.2)

		CASE–A.0 UNCONTROL Max.	CASE–A.0 UNCONTROL Ratio	CASE–A.2 ACC.CONTROL Max.	CASE–A.2 ACC.CONTROL Ratio	CASE–D.2 DIS.CONTROL Max.	CASE–D.2 DIS.CONTROL Ratio
Disp. (cm)	4	1.03	---	0.677	0.734	0.623	0.484
	3	0.840	---	0.528	0.702	0.506	0.482
	2	0.610	---	0.352	0.644	0.406	0.533
	1	0.318	---	0.188	0.660	0.239	0.602
Acc. (gal)	4	148.	---	99.0	0.747	121.	0.655
	3	129.	---	86.0	0.744	120.	0.745
	2	180.	---	100.	0.620	157.	0.699
	1	155.	---	106.	0.764	157.	0.811
Input Acc.(gal)		68.0	---	60.9	---	84.9	---
Added Mass	Disp.(cm)	1.60	---			5.56	
	Acc.(gal)	2260.	---			1210.	

$$RATIO = [\frac{Response(Control)}{Response(Uncontrol)}] \Big/ [\frac{Input(Control)}{Input(Uncontrol)}]$$

6.3 Seismic Wave Input Experiment Results

The maximum response values in case of input with the time axis of El Centro NS wave reduced to 1/3 are compared for the three cases and given in Table 3.

Since a long-period component of approximately 28 sec. was seen in the DVA mass displacement waveform in the acceleration evaluation control, in subsequent processes, all long-period compoents of 10 sec. or more were cut out.

When considering based on the response ratio with no-control, and compare the stroke of the DVA required for acceleration decrease of about the same degress, it will be permissible for the stroke to be much smaller in acceleration evaluation control than in displacement evaluation control. In case of placing importance on restricting conditions of stroke, it may be said that acceleration evaluation control is a method of good efficiency.

Fig.9 shows the waveforms of the second story as an example of comparing the response waveforms of the structure in acceleration evaluation and displacement evalution control experiments.

In Fig.10, the responses of the DVA mass are similarly compared, and the difference in responses of the DVA mass according to the difference in evaluation function is clearly seen.

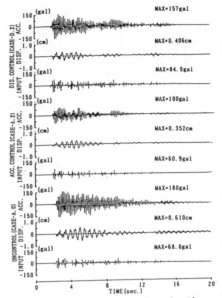

Fig.9 Response displacements and accelerations
(CASE A.0 , CASE A.2 and CASE D.2)
(El Centro Earthquake 1/3 time scale input)

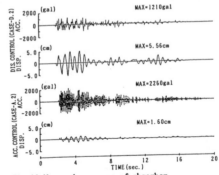

Fig.10 Mesured response of absorber
(CASE A.2 and CASE D.2)
(El Centro Earthquake 1/3 time scale input)

7. CONCLUSION

The method of optimum design of an active dynamic vibration absorber using an AC servo motor and a ball screw has been presented, along with which the effectiveness has been verified by means of vibration tests. In the past, mass ratios of 1 to 2 % were mainly used in studies of active dynamic vibration absorbers, but it has been ascertained that even with about 0.5 %, equivalent or better vibration control performance can be achieved if optimum control considering high-order modes is realized, and that analytical results can explain experimental results well and there is ample possibility for them to be applied to actual structures.

On the other hand, in active control experiments by acceleration evaluation control, some problems about sensors were experienced and experiments giving large control forces could not be conducted. But the good efficiency of acceleration evalution control was ascertained, such as the feasibility of reducing acceleration to the same degree with small strokes compared with relative displacement evaluation control. It is planned for sensors to be improved in the furure, and experiments giving even larger control forces to be conducted.

AKNOWLEDGEMENTS

This development research has been a joint undertaking of the Shimogo Laboratory and Yoshida Laboratory of the Faculty of Science and Engineering, Keio University, Tokiko Corporation, and Obayashi Corporation, and the authors wish to express their gratitude to the persons concerned at these institutions.

The authors also express their gratitude to Dr. Hidekazu Nishimura of Chiba University, formerly of Keio University, for his valuable advice regarding analytical aspects.

REFERENCES

(1) H.Nishimura, K.Yoshida, and T.Shimogo, "An Optimal Active Dynamic Vibration Absorber for Multi-degrees-of-Freedom Systems (A Control Using a Reduced-Order Model and Output Feedback Controls)," Trans. of JSME(C), Vol.54, No.508, August 1988, pp 2948-2956.

(2) H.Nishimura, K.Yoshida, and T.Shimogo, " Optimal Active Dynamic Vibration Absorber for Multi-degree-of-Freedom Systems (Feedback and Feedforward Control Using a Kalman Filter)," Trans. of JSME (C), Vol.55, No.517, September 1989, pp 2321-2329.

Optimal Control of Distributed Parameter
Systems under Resonant and Unstable Loading

by

Iradj G. Tadjbakhsh
and Yuan—an Su

Rensselaer Polytechnic Institute
Troy, New York 12181

Abstract. Vibrations of linear, conservative distributed—parameter systems can be suppressed by a set of discrete actuators and an optimal control algorithm which minimizes the current value of a positive—definite, time—dependent objective function. The control procedure depends on the initial data and is independent of the ultimate outcome.

The control algorithm that is developed has a limited time interval of applicability about the arbitrary initial point. Outside this interval, negative damping and reduced stiffness may arise. However, the arbitrariness of the initial point allow re—initiation of the end—point and thus provides a means of continuation of control.

Using variational methods and the general Duhamel integral representation of the solution, explicit representation for infinite dimensional gain matrices is obtained. Examples of successful control under conditions of resonant loading and moving loads are carried out. The control algorithm can also be modified to control a beam—column subjected to transverse loads on the span the span before the beam is buckled. Due to loss of controller cannot suppress the vibrations of a buckled beam. The question of the effect of control force spillover into higher uncontrolled modes of the system is considered.

The Control Problem. Consider a flexible elastic medium such as a plate, a rod, a column, a membrane or a cable. Let $z(\underset{\sim}{r},t)$ denote the deflection of the structure with $\underset{\sim}{r}$ and t respectively representing the position of a material point in a domain D and the time. Assume n actuators are at locations $\underset{\sim}{r}_i$, $i = 1,2,3,...,n$ with control forces u_i (t). Motion of the controlled structure will be described by

$$m(\underset{\sim}{r})\, \ddot{z} + Lz = f(\underset{\sim}{r},t) + \sum_{i=1}^{n} u_i\,(t)\, \delta(\underset{\sim}{r} - \underset{\sim}{r}_i),\ \underset{\sim}{r}\epsilon D,\ t > t_o \qquad (1)$$

where $m(\underset{\sim}{r})$ is the mass distribution, $f(\underset{\sim}{r},t)$ is the external excitation, δ is the Dirac delta function and L is a self–adjoint spatial operator with self–adjoint boundary conditions.

The initial time t_0 is any arbitrary point on the time scale at which the initial conditions

$$z_0(\underset{\sim}{r}) = z(\underset{\sim}{r},t_0) \quad , \quad \dot{z}_0(\underset{\sim}{r}) = \dot{z}(\underset{\sim}{r},t_0) \tag{2}$$

are known. In (1) and (2) a dot denotes time derivative.

Associated with the above system is the positive definite potential energy $U(t)$ whose variation δU due to a small variation $\delta z(\underset{\sim}{r},t)$ is given by

$$\delta U = \int_D Lz \, \delta z \, dD \tag{3}$$

A positive performance index J is defined for the current value of time and consists of the kinetic energy, potential energy and the control effort

$$J(t) = Q_1 \left(\int_D \frac{1}{2} m \, \dot{z}^2 \, dD \right) + Q_2 U + \frac{1}{2} \sum_{i=1}^{n} R_i u_i^2 \tag{4}$$

The positive functions $Q_1(t)$, $Q_2(t)$ and $R_i(t)$ determine the level of the effectiveness of the control criteria and can be assigned arbitrary values by the designer. The control forces u_i will be determined such that $J(t)$ is a minimum subject to the constraint of the equations of motion $(1) - (2)$. This minimization implies that the system will acquire a state as near its state of rest $z = \dot{z} = 0$ as it is possible with limited control effort. Using the Duhamel Integral representation for the modal amplitudes $\underset{\sim}{a}(t)$ of z and employing variational methods the control force is determined to be

$$\underset{\sim}{u} = - Q_1 \underset{\sim}{D} \, \underset{\sim}{a}_c - Q_2 \underset{\sim}{E} \, \underset{\sim}{a}_c \tag{5}$$

where

$$\underset{\sim}{D} = \underset{\sim}{R}^{-1} \underset{\sim}{B}_c^T \underset{\sim}{S}_c \, (t–t_0), \quad \underset{\sim}{E} = \underset{\sim}{R}^{-1} \underset{\sim}{B}_c^T \, [\underset{\sim}{I}_c - \underset{\sim}{C}_c \, (t–t_0)] \tag{6}$$

and $\underset{\sim}{S}$ and $\underset{\sim}{C}$ are diagonal matrices of the modal harmonics given by

p 223

$$\underset{\sim}{S} = \text{diag}\,[\omega_j^{-1}\,\sin\,\omega_j t]\ ,\ j = 1,2,3$$

$$\underset{\sim}{C} = \underset{\sim}{\dot{S}} = \text{diag}\,[\cos\,\omega_j t]\ ,\ j = 1,2,3$$

The subscript c denotes the truncation of the infinite dimensional matrices and vectors corresponding to the modes that are being controlled and ω_j are the modal frequencies. Assuming that the effect of time delay due to computation on the system is small enough to be negligible, the behavior of the controlled modes are determined to be

$$\underset{\sim}{\ddot{a}}_c + Q_1 \underset{\sim}{B}_c\, \underset{\sim}{D}\, \underset{\sim}{\dot{a}}_c + (\underset{\sim}{\Omega}_c^2 + Q_2 \underset{\sim}{B}_c\, \underset{\sim}{E})\, \underset{\sim}{a}_c = \underset{\sim}{f}_c$$

The residual modes are subject to control spillover and are determined from

$$\underset{\sim}{\ddot{a}}_r + \underset{\sim}{\Omega}_r^2\, \underset{\sim}{a}_r = \underset{\sim}{f}_r - Q_1 \underset{\sim}{B}_r\, \underset{\sim}{D}\, \underset{\sim}{\dot{a}}_c - Q_2 \underset{\sim}{B}_r\, \underset{\sim}{E}\, \underset{\sim}{a}_c$$

The the control interval, i.e., $t - t_0$ should not exceed one–half of the period of the highest controlled mode, i.e.,

$$(t - t_0) \le \frac{\pi}{\omega_{N_c}}$$

Control of A Simply Supported Beam. Consider a uniform beam hinged at both ends and controlled by four equally spaced actuators. The lowest ten modes of the beam constitute the controlled and the uncontrolled modes.

In the first example, the beam is subjected to a unit cyclic, concentrated load resonant with the first mode of the beam acting at the midspan. The control intervals are 0.01, 0.08 and 0.80 seconds and only the first mode is subjected to control. Control of additional modes is not called for as they are not being excited. The results that are presented refer to the deflection at midspan of the beam as shown in Fig. 1. It can be seen that, unlike the uncontrolled beam the response remains bounded.

Control of Simply Supported Beam–Columns. Beam columns are members that are subjected to both bending and axial compression. They will be unstable when the axial compression reaches certain critical values. The system is subjected to an impulse at midspan. The control intervals are set to 0.01 and 0.05 seconds to compare results. The behavior of the beam–column is assumed to be adequately represented by its first ten modes and of these the lowest five are controlled. Fig. 2 shows the response of the system over a 10 seconds interval. The axial compressive load ratio $p = P/P_E$ is set to 0.9. The deflection of the beam remains bounded and is accompanied by control force spilling over into the higher modes.

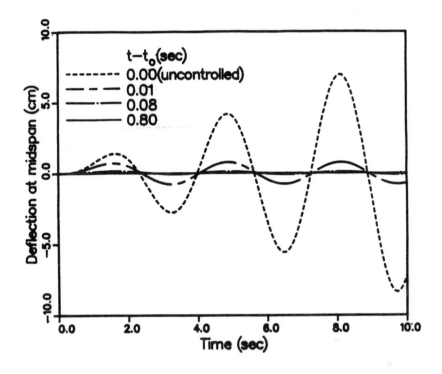

Fig. 1   The midspan deflection of the beam subjected to an unit periodic load
          resonant with the first fundamental frequency of the beam, $t-t_o = 0.01$,

          0.08, 0.80 seconds.

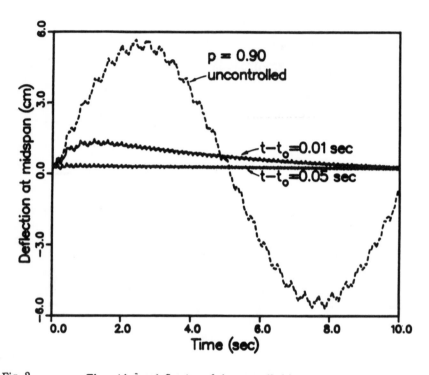

Fig. 2   The midspan deflection of the controlled beam–column subjected to an unit impulse at midspan, p = 0.9, t–t$_o$ = 0.01, 0.05 seconds.

U.S. National Workshop on Structural Control Research
University of Southern California
October 25-26, 1990

EXPERIMENTAL STUDY ON ACTIVE CONTROL
OF CABLE-STAYED BRIDGES

Pennung WARNITCHAI, Benito M. PACHECO and Yozo FUJINO
Department of Civil Engineering, University of Tokyo

Background: Active tendon control seems ideal for the suppression of vibrations in a cable-stayed bridge since the stay cables can serve as active tendons. The feasibility of suppressing wind-induced vibrations of the bridge girder by regulating the cable tension forces through hydraulic servomechanisms, was studied theoretically by Yang and Giannopolous [1]. In the present experimental study, it is emphasized that the cables themselves may in fact vibrate locally, and may strongly interact with the girder under certain conditions [2,3]. Active stiffness control [4], which is another scheme using the same hardware as active tendon control, is also investigated. Active stiffness control focuses on the cable vibrations, while active tendon control considers mainly the vibrations of the girder.

Structural modeling and basic design of control algorithms: 'Global' vibration refers to the motion of the girder, pylon and cables as one assemblage, with the cables behaving like elastic massless tendons; this was considered in [1]. 'Local' vibration, on the other hand, refers to transverse oscillation of a cable. The interaction between these two types of vibration can be strong when there is resonance among the global and local modes; and this coupling must be recognized when designing the control. Both global and local mode shapes, which are assumed to be classical (real), are therefore used as generalized coordinates in deriving the reduced-order structural model [2]. The model includes linear and nonlinear interaction among global and local modes. The nonlinearity is geometric and is attributed to the finite motion of the cables.

Two control schemes are used [2]. The first is *active tendon control,* which is designed by focusing on a global mode. The actuator motion is controlled to be proportional to the velocity of the mode. By assigning a high control gain, the structural response and actuator motion can be kept within small amplitude. In this algorithm and the next, it is deemed prudent to limit the amplitude of the control-generated additional tension in the cable to within a few percent of the static tension.

The second scheme, which is *active stiffness control*, is designed by focusing on one local mode at a time. The mechanism of active stiffness control is basically a reversed parametric excitation, i.e., in case of harmonic motion, the axial displacement causes a variation in cable tension at a frequency that is twice the frequency of the local motion that is being controlled, and with the proper 'phase'. In the case of nearly harmonic motion with slightly and slowly varying frequency or amplitude, these requirements in frequency and phase, plus a constraint to keep the amplitude of actuator motion practically constant, are implemented by a nonlinear feedback law.

Experimental set-up: A simplified scale model with dynamic similarity to an actual bridge is used [2,3]. Fig.1 shows schematically the set-up for active tendon control of a global mode. The scale model comprises a two-meter steel beam, a steel stay wire (cable), a piezoelectric actuator at the upper anchorage of the cable, an analog control unit that allows manual adjustment of gain or phase, optical position sensors for the cable vibration, and strain gages for the beam. A gap sensor is installed near the upper cable anchorage to measure the actual motion of the actuator. An oscilloscope allows visual verification of the phase relationship between the vibration signal and actuator signal. The manual control of phase allows *ad hoc* compensation for such effects as time-delay and parameter uncertainties, which must be studied in later experiments. For further simplicity in the present study, only harmonic excitation is considered. Electromagnetic exciters are used to apply harmonic forces to either the cable or the beam.

Results: Several structural configurations of the cable-stayed beam model were studied. The configuration was varied by tuning or detuning the static tension in the cable. To test the effectiveness and predictability of active tendon control, the beam was excited vertically near the natural frequency of the *second* global in-plane mode (about 19.8 Hz), which is the mode to be controlled. Active tendon control proved to be generally very effective when this mode is detuned from the cable.

In one case where the proportion of three natural frequencies (second global in-plane: first global out-of-plane: first local out-of-plane) was nearly 2:1:1 and therefore was prone to autoparametric resonance as discussed in [3], active tendon control of the second global in-plane mode was again effective, in fact so effective that the autoparametric effect on the two out-of-plane modes was completely suppressed.

However, instability due to observation and control spillover tended to occur when the assigned control gain was very high. In the case considered, the natural frequencies of first global in-plane, second global in-plane, and second local in-plane modes were 6.7, 19.8, and 17.2 Hz, respectively. Two principal modes of this configuration are rather closely spaced and are combinations of second global and second local in-plane modes, which are weakly coupled. One principal mode, at about 17.04 Hz, is local-dominated, and the other, at about

19.82 Hz, is global-dominated; the undamped mode shapes are shown schematically in Fig.2.

When the control gain was set at a high level, instability due to spillover from the *first* global in-plane mode (about 6.7 Hz) was observed; this could be removed by using a band-pass filter, instead of low-pass filter, that was placed in the control loop before the phase shifter. When the control gain was further increased, this time the *local-dominated* in-plane mode (about 17.04 Hz), caused instability; but the observation spillover could not be removed by ordinary band-pass filter because the frequency was too close to the *global-dominated* in-plane mode (about 19.82 Hz) that was being controlled.

In the final case that was considered for active tendon control, the local and global modes were strongly coupled and hence the frequencies of the local-dominated and global-dominated in-plane principal modes were more closely spaced (about 19.25 and 19.90 Hz). Accordingly, very high control gain had to be avoided in order to prevent instability. The active damping algorithm was ineffective in suppressing vibrations in the local-dominated mode around 19.25 Hz (see Fig.3). Among the possible causes of the ineffectiveness are: (1) although the cable itself is very light, the effective mass of the local-dominated mode is about nine times that of the global-dominated; and (2) the phase relationship prescribed by the algorithm does not recognize the modal distortion that was in fact observed in the experiment, e.g., there was a significant phase difference between the motions of two locations in the beam (complex mode).

For testing active stiffness control, the beam was temporarily clamped and the cable was tied with a small thread in the vertical plane, in order that pure local out-of-plane vibrations in a single mode could be excited (see Fig.4). Active stiffness control proved to be very effective, even as the cable exhibited a moderate hardening due to geometric nonlinearity before the application of control (see the bent frequency response curve in Fig.5).

In cable-stayed bridges, the two types of vibration require different control schemes. In active tendon control, the present feasibility study indicates that modifications are necessary when one or more modes are close to the mode being controlled. When a local-dominated mode is among the closely-spaced modes, and this has to be suppressed, active stiffness control may be the appropriate algorithm. Both control schemes are discussed in detail in Refs. 5 and 6.

References: [1] Yang, J. N. and Giannopolous, F., *J. Eng. Mech. Div.,* ASCE, Vol.105, No.4, pp.677-694, August 1979 and No.5, pp.795-810, October 1979. [2] Warnitchai, P., *Doctoral Dissertation,* Univ. of Tokyo, September 1990. [3] Pacheco, B.M., Fujino, Y. and Warnitchai, P., 'An Experimental and Analytical Study of Autoparametric Resonance in a 3DOF Model of Cable-Stayed Beam,' submitted for publication in *Nonlinear Dynamics,* Kluwer. [4] Chen, J.-C., *J. Spacecraft and Rockets,* AIAA83-0344, Vol.21, No.5, pp.463-467. [5] Warnitchai, P., Fujino, Y. and Pacheco, B.M., 'Active Tendon Control of Cable-Stayed Bridges,' (in preparation). [6] Fujino, Y., Warnitchai, P. and Pacheco, B.M., 'Active Stiffness Control of Cable Vibration,' (in preparation).

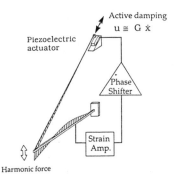

Active damping
$u \cong G \dot{x}$

Piezoelectric
actuator

Phase
Shifter

Strain
Amp.

Harmonic force

Fig.1 Set-up for active tendon control

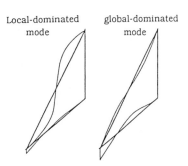

Local-dominated
mode

global-dominated
mode

Fig.2 Closely spaced principal modes

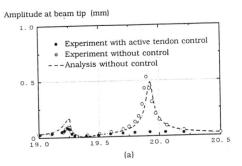

Amplitude at beam tip (mm)

• Experiment with active tendon control
○ Experiment without control
‑ ‑ ‑ Analysis without control

(a)

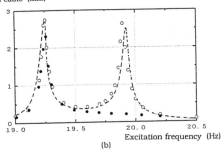

Amplitude at quarter point
of cable (mm)

Excitation frequency (Hz)

(b)

**Fig.3 Frequency response with and without
active tendon control**

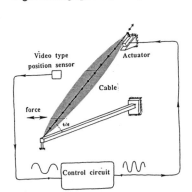

Video type
position sensor

Actuator

Cable

force

tie

Control circuit

Fig.4 Set-up for active stiffness control

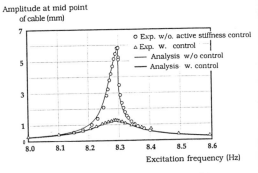

Amplitude at mid point
of cable (mm)

○ Exp. w/o. active stiffness control
△ Exp. w. control
— Analysis w/o control
— Analysis w. control

Excitation frequency (Hz)

**Fig.5 Frequency response with and without
active stiffness control**

p 230

ACTIVE CONTROL OF THE SEISMIC RESPONSE OF STRUCTURES IN THE PRESENCE OF SOIL-STRUCTURE INTERACTION EFFECTS

H. L. Wong[1] and J. E. Luco[2]

ABSTRACT

The effects of soil-structure interaction on the form of the control rule and on the effectiveness of active control of the seismic response of structures are examined. The structure is modeled as a uniform shear beam supported on a rigid foundation embedded in an elastic soil. Active control in the form of an absorbing boundary located at the top of the structure is considered. It is found that the rocking of the foundation changes the form of the control rule. However, the effectiveness of this form of active control is not degraded when soil-structure interaction effects are included.

INTRODUCTION

A common assumption in most studies on active control of the seismic response of structures is that soil-structure interaction effects are small and, in particular, that the rocking motion of the base is negligible. The objective of this study is to remove these assumptions and to consider the seismic response of tall structures subjected to active control when the flexibility of the soil is included in the analysis. In this paper, the structure is modeled as a uniform shear beam supported on a rigid foundation embedded in the soil represented by a uniform viscoelastic half-space (Fig. 1). The seismic excitation is represented in the form of vertically incident SH-waves. The kinematic interaction effects associated with the embedment of the foundation together with the inertial interaction effects result in a base motion that includes translational and rocking response components. The seismic response of the structure including soil-structure interaction effects is modified by use of a control force acting at the top of the structure. The active control strategy used here is based on the work of Vaughan (1968) and von Flotow (1986) in which the energy flow within the structure is modified by controlling the reflection and/or transmission of waves at end points or at joints. In this study, the active control force is selected to simulate an absorbing boundary such that all upward propagating waves are absorbed at the top of the structure and no downward propagating waves are reflected at that point. Applications of this approach to the active control of the seismic response of tall structures in the absence of soil-structure interaction effects have been presented by Mita and Luco (1990a,b) and Luco et al (1990).

FORMULATION OF THE PROBLEM AND SOLUTION

Basic Equations. We consider first the motion of the superstructure for a given translation and rotation of the base. The equation of motion of the structure and the boundary conditions for harmonic vibrations with time dependence $e^{i\omega t}$ are given by

$$u'' + (\omega/\beta_B)^2 u = 0, \qquad 0 < x < H \tag{1}$$

$$u(0) = u_B \tag{2}$$

$$\rho_B A_B \beta_B^2 [u'(H) - \theta_B] = F_T \tag{3}$$

where $u(x)e^{i\omega t}$ is the total translation of the structure with respect to an inertial frame of reference, $u_B e^{i\omega t}$ is the horizontal motion of the base of the structure (top of the rigid foundation), θ_B is the rocking angle of the base, $F_T e^{i\omega t}$ is the active control force applied at the top of the structure, ρ_B and A_B are the density and cross-sectional area of the shear wall, respectively, and $\beta_B = \bar{\beta}_B(1 + 2i\xi_B)^{1/2}$, in which ξ_B is the hysteretic damping ratio in the structure and $\bar{\beta}_B$ is (approximately) the shear wave velocity within the structure. The general solution of Eq. (1) is

$$u(x) = Ae^{-i(\omega x/\beta_D)} + Be^{i(\omega x/\beta_D)} \tag{4}$$

---

[1] Department of Civil Engineering, University of Southern California, Los Angeles, California 90089-2531.

[2] Department of Applied Mechanics and Engineering Sciences, University of California, San Diego, La Jolla, Ca 92093-0411.

in which the first and second terms represent upward and downward propagating waves, respectively.

Structural Response without Absorbing Boundary (Case 1). To evaluate the effectiveness of active control we consider first the response of the structure when no control is provided. In this case, the boundary condition at the top of the structure is given by

$$F_T = \rho_B A_B \beta_B^2 [u'(H) - \theta_B] = 0 \qquad (5)$$

Substitution of the general solution given by Eq. (5) into the boundary conditions given by Eqs. (2) and (5) permits us to determine the unknown coefficients A and B with the result

$$u(x) = \left[\cos\left(\omega x/\beta_B\right) + \tan\left(\omega H/\beta_B\right)\sin\left(\omega x/\beta_B\right)\right] u_B + \left(\beta_B/\omega L\right)\frac{\sin\left(\omega x/\beta_B\right)}{\cos\left(\omega H/\beta_B\right)} L\theta_B \quad . \qquad (6)$$

Structural Response with Absorbing Boundary (Case 2).

If the control force F_T is selected so that an absorbing boundary is obtained at the top of the structure, then no reflected waves are obtained and $B = 0$. In this case, Eqs. (2) and (4) with $B = 0$ lead to

$$u(x) = u_B e^{-i(\omega x/\beta_D)} \qquad . \qquad (7)$$

The required control force F_T is obtained by substitution from Eq. (7) into Eq. (3). The resulting expression is

$$F_T = -\rho_B A_B \beta_B \dot{u}(H) - \rho_B A_B \beta_B^2 \theta_B \qquad (8)$$

where the dot denotes differentiation with respect to time $[\dot{u}(H) = i\omega u(H)$ for harmonic time dependence]. The control rule given by Eq. (8) differs from previous expressions derived under the assumption of negligible rocking (Mita and Luco, 1990a,b) in that it depends not only on $\dot{u}(H)$ but also on θ_B.

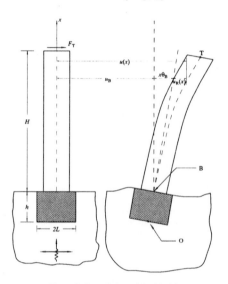

Figure 1. Description of the Model

Structural Response with a Simplified Absorbing Boundary (Case 3). It is of interest to consider the response of the structure when the control rule given by Eq. (8) is replaced by the simpler expression

$$F_T = -\rho_B A_B \beta_B \dot{u}(H) \qquad (9)$$

Eq. (9) corresponds to the form of the control rule when rocking of the base is ignored (Mita and Luco, 1990a,b). In this case, substitution of the general solution, given by Eq. (2), into the boundary conditions given by Eqs. (2) and (3) with F_T given by Eq. (9) results in the solution

$$u(x) = u_B e^{-i(\omega x/\beta_D)} + \theta_B \left[\sin(\omega x/\beta_B)/(\omega/\beta_B)\right] e^{-i(\omega H/\beta_D)} \qquad (10)$$

The solution in this case includes both upward and downward propagating waves in the structure.

Soil-Structure Interaction Equations. It can be shown (Luco and Wong, 1982) that the generalized displacement of the bottom of the foundation $\{\tilde{U}_0\} = (u_0, L\theta_0)^T$ normalized by the half-width L of the foundation is given by

$$\{\tilde{U}_0\} = \left([I] + [C(\omega)]\left([\tilde{K}_{B0}(\omega)] - (a_0^2/\rho_s L^3)[\tilde{M}_0]\right)\right)^{-1}\{\tilde{U}_0^*\} \qquad (11)$$

where $[I]$ is the 2×2 identity matrix, $[C(\omega)]$ is the 2×2 normalized foundation compliance matrix, $[\tilde{M}_0]$ is the normalized mass matrix for the rigid foundation, $a_0 = \omega L/\bar{\beta}_s$ is a dimensionless frequency normalized by L and by the shear wave velocity of the soil $\bar{\beta}_s$, ρ_s is the density of the soil and $\{\tilde{U}_0^*\} = (u_0^*, L\theta_0^*)^T$ is the effective input motion to the foundation. In this study, we assume that the seismic excitation corresponds to vertically incident shear waves characterized by the free-field ground motion on the soil surface $u_g(\omega)e^{i\omega t}$. In this case,

$$u_0^* = S(a_0)u_g \quad , \quad L\theta_0^* = R(a_0)u_g \qquad (12)$$

where $S(a_0)$ and $R(a_0)$ are scattering coefficients which depend on the dimensionless frequency a_0 and on the characteristics of the foundation and the soil. The scattering coefficients and the elements of the compliance matrix (or, of its inverse the impedance matrix) can be obtained from published results (e.g., Mita and Luco, 1989). The 2×2 matrix $[\tilde{K}_{B0}(\omega)]$ appearing in Eq. (11) relates the generalized force $\{\tilde{F}_{B0}\} = (F_{B0}, M_{B0}/L)^T$ that the superstructure exerts on the foundation with the motion $\{\tilde{U}_0\}$ of the bottom of the foundation. The elements of the matrix $[\tilde{K}_{b0}(\omega)]$ are obtained by considering the linear and angular momenta of the superstructure.

Once the motion $\{\tilde{U}_0\}$ of the bottom of the foundation has been obtained by use of Eqs. (11), the motion $\{\tilde{U}_B\} = (u_B, L\theta_B)^T$ at the base of the superstructure, the response $u(x)$ at any point in the structure, the base shear force F_B, the base overturning moment M_B and the required control force F_T can be easily calculated.

NUMERICAL RESULTS

To study the effects of soil-structure interaction on the effectiveness of active control of the seismic response of structures we have considered a simplified model of a 10-storey building founded on soils with different rigidities. The structural model is characterized by the fundamental fixed-base period $T_1 = 1.0$ sec, fixed-base hysteretic damping ratio $\xi_B = 0.02$, velocity of shear waves in the building $\bar{\beta}_B = 150$ m/sec, slenderness ratio $H/L = 3.75$, mass ratio $(\rho_B A_B H/\rho_s L^2 H) = 1.0$ and inertia ratio $(\rho_B I_B H/\rho_s L^5) = 0$. The foundation was modeled as rigid rectangular block of base dimension $2L \times 2L$ embedded to a depth h in the soil. Values of the half-width $L = 10$ m, embedment ratio $h/L = 0.50$, foundation mass ratio $(M_F/\rho_s L^3) = 0.7$ and foundation inertia ratio $(I_F/\rho_s L^5) = 0.35$ were considered. The soil was modeled as a uniform viscoelastic half-space characterized by complex wave velocities $\alpha_s = \bar{\alpha}_s(1 + 2i\xi_\alpha)^{1/2}$ and $\beta_s = \bar{\beta}_s(1 + 2i\xi_\beta)^{1/2}$ for P- and S-waves, respectively. To take advantage of the numerical results presented by Mita and Luco (1989) for the impedance functions and scattering coefficients of square embedded foundations it was assumed that $\bar{\alpha}_s = 2\bar{\beta}_s$ ($\nu \approx 1/3$), $\xi_\alpha = 0.0005$ and $\xi_\beta = 0.001$. Three values of the soil shear wave velocity $\bar{\beta}_s$ corresponding to $\bar{\beta}_s = 1500$ m/sec, 300 m/sec and 150 m/sec were used to represent stiff, intermediate and soft soil conditions.

Numerical results for the 10-storey building model were obtained for three soil conditions and for three cases corresponding to the absence of control (Case 1), control by the absorbing boundary defined by Eq. (8) (Case 2), and control by the simplified absorbing boundary defined by Eq. (9) (Case 3). Results in the frequency domain for intermediate soil conditions $\bar{\beta}_s = 300$ m/sec are shown in Fig. 2. The amplitudes of the transfer functions $|u_T/u_g|$ where $u_T = u(H)$ is the total motion at the top of the structure and u_g is the free-field motion of the ground surface are shown in Fig. 2(a). A first observation is that both the exact (Case 2) and the simplified (Case 3) absorbing boundaries drastically reduce the response and eliminate all resonant behavior. Both absorbing boundaries lead to almost the same response at the top. The results for $|u_T/u_g|$ also indicate that the beneficial effects introduced by active absorbing boundaries are not

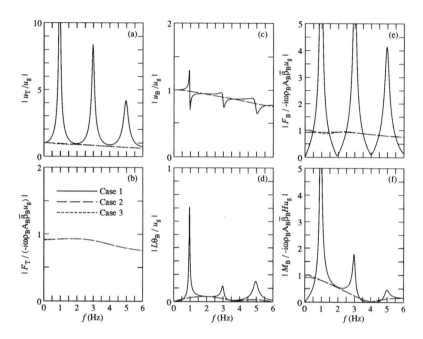

Fig. 2.   Normalized Amplitudes of: (a) Top Translation $|u_T/u_g|$,   (b) Control Force $|F_T/(-i\omega\rho_B A_B\bar\beta_B u_g)|$, (c) Base Translation $|u_B/u_g|$, (d) Base Rotation $|L\theta_B/u_g|$, (e) Base Shear Force $|F_B/(-i\omega\rho_B A_B\bar\beta_B u_g)|$ and (f) Base Overturning Moment $|M_B/(-i\omega\rho_B A_B\bar\beta_B H u_g)|$ for a 10-Storey Building on an Intermediate Soil ($\bar\beta_s = 200$ m/sec). Cases 1, 2 and 3 Correspond, Respectively, to Absence of Control, Control by an Exact Absorbing Boundary and Control by a Simplified Absorbing Boundary.

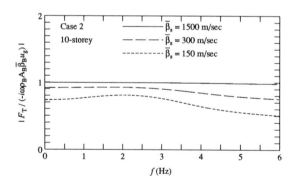

Fig. 3.   Effect of Soil-Structure Interaction in the Amplitude of the Normalized Control Force $|F_T/(-i\omega\rho_B A_B\bar\beta_B u_g)|$ for a 10-Storey Building.

reduced in any way by soil-structure interaction effects. The normalized values for the amplitude of the required control force $|F_T/(-i\omega\rho_B A_B\bar{\beta}_B u_g)|$ shown in Fig. 2(b) indicate that both types of absorbing boundaries lead to almost the same control force. The amplitudes of the transfer functions $|u_B/u_g|$ and $|L\theta_B/u_g|$ for the translational and rocking response at the base of the structure shown in Figs. 2(c) and 2(d), respectively, indicate that the use of absorbing boundaries reduces the inertial interaction effects on the base translation and rotation. In particular, the rocking response is drastically reduced when absorbing boundaries are used. As in the case of the response at the top, both types of absorbing boundaries (Cases 2 and 3) lead to almost the same response at the base. The normalized amplitudes $|F_B/-i\omega\rho_B A_B\bar{\beta}_B u_g|$ and $|M_B/-i\omega\rho_B A_B\bar{\beta}_B H u_g|$ of the base shear force and base overturning moment are shown in Figs. 2(e) and 2(f), respectively. These results indicate that the use of absorbing boundaries strongly reduces the base shear force and the base overturning moment in the vicinity of the characteristic frequencies of the system without control (Case 1). At other frequencies these quantities may be increased by the use of control through absorbing boundaries.

The effects of soil-structure interaction on the amplitude of the normalized control force $F_T/(-i\omega\rho_B A_B\bar{\beta}_B u_g)$ are summarized in Fig. 3. These results indicate that the amplitude of the required control force decreases as the soil becomes softer. It appears that this reduction is mainly associated with kinematic interaction effects and, consequently, will be stronger for larger structures founded on deeper foundations.

CONCLUSIONS

The effects that the interaction between the structure and the soil may have on the possibility of using active control techniques to modify the seismic response of structures have been studied. It has been found that the rocking of the foundation resulting from the kinematic and inertial interaction effects changes the form of the control rule required to obtain an active absorbing boundary at the top of the structure. Active control by means of an absorbing boundary which includes the rocking effects and by a simplified absorbing boundary which ignores the rocking effects result in large reductions of the structural response even when soil-structure interaction effects are included. In fact, the amplitudes of the required control force and of the structural response decrease as the soil becomes softer. The use of control by means of absorbing boundaries also reduces the inertial interaction effects and, in particular, drastically reduces the rocking response of the structure.

ACKNOWLEDGEMENT

The work described here was supported by grants from the Ohsaki Research Institute, Shimizu Corporation to the University of Southern California and the University of California, San Diego.

REFERENCES

Luco, J. E. and H. L. Wong (1982). "Response of Structures to Nonvertically Incident Seismic Waves," *Bull. Seism. Soc. Am.*, Vol. 72, No. 1, 275-302.

Luco, J. E., H. L. Wong and A. Mita (1990). "Active Control of the Seismic Response of Structures by Combined Use of Base Isolation and Absorbing Boundaries," *Earthquake Engineering and Structural Dynamics*, (submitted for publication).

Mita, A. and J. E. Luco (1989). "Impedance Functions and Input Motions for Embedded Square Foundations," *Journal of Geotechnical Engineering*, GED, ASCE, Vol. 115, No. 4, 491-503.

Mita, A. and J. E. Luco (1990a). "Active Vibration Control of a Shear Beam with Variable Cross Section," *Proceedings of the 1990 Dynamics and Design Conference, Japan Society of Mechanical Engineers* (Kawazaki, Japan, July 9-12), 276-279.

Mita, A. and J. E. Luco (1990b). "New Active Control Strategy for Tall Buildings," *Proceedings Eighth Japan Earthquake Engineering Symposium* (submitted for publication).

Vaughan, D. R. (1968). "Application of Distributed Parameter Concept to Dynamic Analysis and Control of Bending Vibrations," *Journal of Basic Engineering*, June, 157-166.

von Flotow, A. H. (1986). "Travelling Wave Control of Large Spacecraft Structures," *Journal of Guidance, Control and Dynamics*, Vol. 9, No. 4, 462-468.

CHAOTIC DYNAMICS AND STRUCTURAL CONTROL[1]

C. Y. Yang[2]   Alexander H-D. Cheng
Professor   Associate Professor

Civil Engineering Department
University of Delaware
Newark, DE 19716

ABSTRACT

Research in the area of control of traditional civil engineering structures
(buildings, and bridges) and large space structures has a short and recent
history of about twenty years. In 1972 Yao [1] presented the pioneer paper on
concept of structural control, in an attempt to stimulate interest among
structural engineers in the applications of control theory to the design of
civil engineering structures. Since then increasing research activities have
been reported in two International Symposiums on Structural Control, the first
one in June, 1979 [2] and the second in July 1985 [3]. More recently, text
books on Control of Structures have been published by Leipholz and Abdel-Rohman
[4] and Meirovitch [5]. For the current state of the art there is a review on
active control of civil engineering structures in 1988 by Yang and Soong [6].

In the mean time, structural control engineering began to appear in
practice as reported by Rosenbaum [7] in 1990 on active dampers installed at
the Citicorp Building in New York City and computer-activated counterweights
installed on the top floor of the Kyobashi Seiwa building in Japan. Moreover
laboratory experimental research has been carried out to verify analytical
structural control methods and results by Dehghauyar, et al. [8] and Soong, et
al. [9].

One very important sub-area in structural control which has not been
investigated thoroughly is the difficult topic of nonlinear structural control.
The recent research in space exploration stimulates a study on nonlinear control
of flexible structures by Masri, et al. in 1982 [10] and Dehghanyan, et al. in
1985 [8]. A team of earthquake structural engineering researchers, Reinhorn and
Manolis, [11] investigated the control of inelastic structures in 1987. The
importance of nonlinear structural control has long been recognized because
strong oscillations of structures almost always involves nonlinearities from
system geometry or system material behavior and the control of these severe
oscillations, of course, directly relates to the satisfactory and safe operation
of structures. It is also well known that nonlinear structural oscillations,

---

[1]Abstract submitted to the U.S. National Workshop on Structural Control
Research, University of Southern California, Los Angeles, California
90089-2531, Oct. 25-26, 1990.

[2]Workshop presenter and corresponding author.

controlled or uncontrolled, is difficult to analyze mathematically and even numerically. Consequently our understanding in nonlinear structural control is quite limited. This difficult situation seems to come to a breakthrough at the present time on account of the new approach and new discoveries in nonlinear dynamics currently known as chaotic dynamics or simply chaos.

Chaotic dynamics has emerged as a new exciting and sometimes confusing field of engineering research and education in the last decade. The most enthusiastic advocated in the field have used such strong words as, "Chaos is a new revolution in science," and "engineering in the 1990's" [12-15]. Others may feel that "chaos is nothing new but a subset of stochastic nonlinear dynamics."

The new science of chaos began at the discovery of the "butterfly effect" of Lorenz in 1963 in his study of long time prediction of weather. Using a simple but nonlinear system of three first order differential equations of three variables for fluid velocity amplitude and temperature distributions, Lorenz found that long time weather prediction was impossible because a negligible disturbance in the initial conditions (butterfly effect) would cause drastic and practically unpredictable changes in the output for the state of weather. The phenomenon is now known as chaos due to the sensitive dependence of the nonlinear response to initial conditions. The most fascinating discovery from the Lorenz's weather system, besides the butterfly effect which seems to be negative and discouraging to scientists, is the structure and order of the distribution of the state of the weather in a geometric form, the so-called Lorenz strange attractor. Lorenz's attractor is a three dimensional trajectory plot of the state of the weather with time as the implicit independent variable. With negligible difference of the initial conditions, the trajectories of two different weather predictions are drastically different, but the long time trajectories in the three dimensional state space are attracted to a geometric form, the chaotic or strange attractor, which is found to be unique, very orderly and accurately reproducible.

Since 1963, numerous discoveries of chaos have been reported in physics [17], mathematics [18,19], physiology [20], biology [21], and engineering [13,14,22]. This wide range of discipline involved is not surprising because of the many unsolved significant nonlinear problems which had been left aside until the recent computer revolution and the awareness of chaos. Focusing our attention now to the specific area of structural dynamics, chaotic vibrations are reported recently by Poddar, Moon and Mukherjee [23], Holmes [24], Dowell and Pezeshki [25], Reinhall, Caughey, and Storti [26], Kapitaniak [27], Tongue [28] and Ueda [29-31]. A nonlinear structural system governed by the well known Duffing's equation with hard or soft spring is central to most of the above investigations.

The extension of research in chaotic dynamics to feedback structural control, with the concurrent advancements of chaotic dynamics and structural control well underway, seems quite natural. Indeed Ueda published his study of chaotic control of a nonlinear electrical system in 1978 [32] during the period of his famous work on the chaotic dynamics of the Duffing system [29-31]. Holmes [33,34] and Moon [14] reported their chaotic nonlinear control research after a series of analytical and experimental work on chaotic dynamics [35-37].

p 237

In May, 1990, we reported our studies on the chaotic and stochastic dynamics for a nonlinear structural system with hysteresis and degradant [38]. Parallel to this work, we are currently studying two simple nonlinear structural systems with feedback control, one nonlinear system with linear feedback and the other with nonlinear feedback. Our focus lies first on the identification of chaotic structural oscillations, under a deterministic, steady state, and harmonic excitation. We anticipate no difficulty in chaos identification because of the strong nonlinearities in the system and of prior experience from Ueda [32], Holmes [33,34], Moon [14] and our own recent work [38]. Once a clear chaotic oscillation is found, our second step will be its quantitative characterization by the established methodology including the Liapuhov exponent spectrum, fractal dimension, power spectral density function (PSD) and probability density function (PDF). In particular, our attention will focus on the PSD and PDF because they are directly useful in the damage and reliability assessment of the structural system [39], which is of course the primary interest of structural engineers. The third phase of our work is the study of global behavior of the controlled structural system, when some parameters of the system are changed. Our primary concern on the change of global behavior lies in the occurrence and disappearance of chaotic oscillations, the so-called global bifurcation behavior of the controlled structural system. Our last phase is to extend the long term steady state structural response to include the transient part which is most important in the area of earthquake structural engineering research.

REFERENCE

[1] Yao, James T.P. "Concept of Structural Control," Proceeding of ASCE, Structural Div., Vol. 98, No. ST7, July, 1972.

[2] Leipholz, H. H. (Ed.) "Structural Control," Proceedings of the International IUTAM Symposium on Structural Control, University of Waterloo, Ontario, Canada, June 4-7, 1979, North Holland Publishing Company.

[3] Leipholz, H. H. (Ed.) Second International Symposium on Structural Control, University of Waterloo, Ontario, Canada, July 15-17, 1985.

[4] Leipholz, H. H. and Abdel-Rothman Control of Structures, Martinus Nijhoff Publishers, 1986.

[5] Meirovitch, Leonard Dynamics and Control, John Wiley and Sons, 1985.

[6] Yang, J. N. nad T. T. Soong "Recent Advances in Active Control of Civil Engineering Structures," Probabilistic Engineering Mechanics, Vol. 3, No. 4, 1988, pp. 179-188.

[7] Rosenbaum, David B. and Naoaki Usui "Active Dampers Create a Stir," Engineering News Record, Feb. 8, 1990, pp. 39-40.

[8] Dehghanyar, T. J., S. F. Masri, R. K. Miller, G. A. Bekey and T. K. Caughey "Sub-Optimal Control of Non-linear Flexible Space Structures," Proceedings NASA-JPL Workshop on Identification and Control of Flexible Space Structures, San Diego, CA, June 4-6, 1985.

p 238

[26] Reinhall, P. G., Caughey, T. K. and Storti, D. W., "Order and chaos in a discrete Duffing oscillator: Implications on numerical integration," J. Appl. Mech., ASME, 56, 162-167, 1989.

[27] Kapitaniak, T., Chaos in Systems with Noise, World Scientific Pub., Singapore, 1988.

[28] Tongue, B. H., "Characteristics of numerical simulations of chaotic systems," J. Appl. Mech., ASME, 54, 695-699, 1987.

[29] Ueda, Y., "Random phenomena resulting from nonlinearity in the system described by Duffing's equation," Int. J. Non-linear Mech., 20, 481-491, 1985.

[30] Ueda, Y., "Steady motions exhibited by Duffing's equation: A picture book of regular and chaotic motions," in New Approaches to Nonlinear Problems in Dynamics, ed. P. Holmes, SIAM, 1980.

[31] Ueda, Y., "Randomly transitional phenomenon in the system governed by Duffing's equation," J. of Statistical Phys., 20, 181-196, 1979.

[32] Ueda, Y., H. Doumoto and K. Nobumoto, "An Example of Random Oscillations in Three-Order Self Restoring System," Proc. Electric and Electronic Communications Joint Meeting, Kansai District, Japan, Oct., 1978.

[33] Holmes, P. J. and F. C. Moon, "Strange Attractors and Chaos in Nonlinear Mechanics," Journal Applied Mechanics, ASME, Dec., 1983, Vol. 50/1021-1032.

[34] Holmes, P. J., "Dynamics of a Nonlinear Oscillator with Feedback Control: Local Analysis," Journal of Dynamics, Measurement and Control, ASME, June, 1985, Vol. 107/159-165.

[35] Guckenheimer, J. and P. J. Holmes, Nonlinear Oscillations, Dynamical Systems, and Bifurcations of Vector Field, Springer Verlag, New York, 1983.

[36] Holmes, P. J., "A Nonlinear Oscillator with a Strange Attractor," Philosophical Transactions, Royal Soc. London, Vol. 292 N1394, Oct., 1979, pp. 419-448.

[37] Moon, F. C., "Experiments on Chaotic Motions of a Forced Nonlinear Oscillator: Strange Attractors, Journal Applied Mechanics, ASME, Vol. 47, 1980, pp. 638-644.

[38] Yang, C. Y., A. H-D. Cheng and V. Roy, "Chaotic and Stochastic Dynamics for a Nonlinear Structure System with Hysteresis and Degradation," Second International Conference on Stochastic Structural Dynamics, Boca Raton, FL, May, 1990.

[39] Yang, C. Y. Random Vibration of Structures, Wiley Interscience Pub. Co., 1986.

[9] Soong, T. T., A. M. Reinforn and J. N. Yang "A Standard Model for Structural Control Experiments and Some Experimental Results," Second Symposium on Structural Control, Universiyt of Waterloo, Ontario, Canada, July 15-17, 1985 in Structural Control, (Ed. H. H. Leipholz), Martinus Nijhoff Pub. Co., 1987, pp. 669-693.

[10] Masri, S. F., G. A. Bekey and T. K. Caughey "On-Line Control fo Nonlinear Flexible Structures," Journal of Applied Mechanics, Vol. 49, Dec., 1982, pp. 877-884.

[11] Reinhorn, A. M. and G. D. Manolis "Active Control of Inelastic Structures," Journal of Engineering Mechanics, ASCE, March, 1987, Vol. 113(3), pp. 315-333.

[12] Gleick, J., Chaos, Penguin Books, 1987.

[13] Thompson, J. M. T. and Stewart, H. B., Nonlinear Dynamics and Chaos, Wiley, 1986.

[14] Moon, F. C. Chaotic Vibrations, Wiley Interscience, 1987.

[15] Goldstein, G. and Moon, F., "Coming to terms with chaos," Mechanical Engineering Magazine, ASME, January, 1990.

[16] Lorenz, E. N., "Deterministic nonperiodic flow," J. Atmospheric Sci., 20, 130-141, 1963.

[17] Hao, B. L., Chaos, World Scientific Pub., Singapore, 1984.

[18] Holmes, P. J., New Approaches to Nonlinear Problems in Dynamics, (ed.), SIAM, Proceedings of a Symposium, 1980.

[19] Feigenbaum, M. J., "Quantitative universality for a class of nonlinear transformation," J. Statistical Phys., 19, 25-52, 1978.

[20] Goldberger, A. L. Bhargava, V. and West, B. J., "Nonlinear dynamics of the heartbeat," Physica, 17D, 207-214, 1985.

[21] May, R. M., "Simple mathematical models with very complicated dynamics," Nature, 55, 184-189, 1988.

[22] Berges, P., Pomeau, Y. and Vidal, C., Order and Chaos, Toward a Deterministic Approach to Turbulence, Wiley Interscience, 1987.

[23] Poddar, B., Moon, F. C. and Mukherjee, S., "Chaotic motion of an elastic-plastic beam," J. Appl. Mech., ASME,

[24] Holmes, P. J., "Nonlinear oscillator with a strange attractor," Phil, Trans., Roy. Soc. London, 292A, 419-448, 1979.

[25] Dowell, E. H. and Pezeshki, C., "On the understanding of chaos in Duffing's equation including a comparison with experiments," J. Appl. Mech., ASME, 53, 6-10, 1986.

p 240

ASEISMIC HYBRID CONTROL SYSTEMS

by

J.N. Yang, Z. Li and A. Danielians
Department of Civil, Mechanical
and Environmental Engineering
The George Washington University
Washington, D.C. 20052

EXTENDED ABSTRACT

In recent years, considerable progress has been made in the area of aseismic protective systems for civil engineering struc- tures. Aseismic protective systems, in general, consist of two catagories; namely, passive protective systems and active protec- tive systems. The active protective system differs from the passive one in that it requires the supply of external power to counter the motion of the structure to be protected.

The horizontal components of the earthquake ground motions are the most damaging to the building. An important class of passive aseismic protective systems is the base isolation system, which is able to reduce the horizontal seismic forces transmitted to the structure. Extensive theoretical and experimental research has been carried out for base isolation systems. The application of active control systems to building structures which are sub- jected to strong earthquakes and other natural hazards has become an area of considerable interest both theoretically and experi- mentally in the last decade. Recently, a significant progress has been made in active control of civil engineering structures.

While passive base isolation systems are effective for pro- tecting seismic-excited buildings, there are limitations. Passive systems are limited to low-rise buildings, because for tall buildings, uplift forces may be generated in the isolation system leading to an instability failure. Furthermore, in some base isolation systems, such as lead-core elastomeric bearings, in- elastic or permanent deformation may accumulate after each earth- quake episode. Thus, the passive protective system alone is not sufficiently proven for the protection of seismic-excited tall buildings.

On the other hand, when an active control system is used alone as a primary aseismic protective system for tall buildings, the required active control force and force rate to be provided by the external power source may be very large. Hence, a large or powerful active control system may be needed. For the installation of a large active control system with large stand-by energy sources, the issues of cost, reliability and practicality remain to be resolved.

In our research program, we propose two new types of hybrid control systems for seismic-excited tall buildings [Refs. 1-4]. These hybrid systems consist of a base isolation system, such as elastomeric bearings or sliding systems, connected to either a passive or active mass damper. With such hybrid systems, the advantage of the base isolation system, whose ability to drastically reduce the horizontal motion of the building, is preserved, whereas its safety and integrity are protected by either the passive or active mass damper. The performance of these two proposed hybrid control systems is investigated, evaluated and compared with that of an active control system for a twenty-story building and a five-story building model subjected to a strong earthquake. It is shown that the proposed hybrid control systems are very effective in reducing the response of building structures under strong earthquake excitations and that they are more effective and advantageous than the application of an active control system alone. Likewise, the practical implementation of such hybrid protective systems is much easier than that of an active control system alone [Refs. 1-4].

Under strong earthquake excitations, tall buildings may undergo significant lateral displacements. During a lateral motion, the gravitational load of the building results in an overturning moment. The effect of such an overturning moment is referred to as the P-delta effect. For well designed building structures with small lateral displacement, the P-delta effect is usually of the second order and it may be negligible. However, for buildings implemented by a base isolation system, the lateral displacement of the base isolation system may be significant and hence the P-delta effect may be important. The P-delta effect on the dynamic response of buildings implemented by the proposed hybrid protective systems is also investigated. It is shown that the P-delta effect should be accounted for in the analysis of building structures implemented by the hybrid protective systems [Refs. 2-4].

Recent experimental demonstrations for the application of aseismic control systems to scaled building structures indicate some difficulties involved in the measurement of floor displacements. The main reason is that during earthquake ground motions, both the building and the ground are moving so that there is no absolute reference for the determination of the floor displacement. This is particularly critical for practical implementations of active control systems to full-scale buildings for earthquake hazard mitigations. Laboratory experiments further indicate that the floor displacement response obtained by numerically integrating the velocity measurement differs significantly from the actual floor displacement due to (i) noise pollutions and (ii) error accumulations resulting from numerical integrations. Unfortunately, available optimal control theories require measurements of displacement and velocity responses of the building structure. Since the measurements of acceleration and velocity of the structural response are much easier, it is highly desirable to use acceleration and velocity sensors rather than displacement sensors.

An optimal control theory is proposed in this research program utilizing the acceleration measurements rather than the displacement measurements [Refs. 5-7]. This optimal control law is developed based on the instantaneous optimal control theories developed by Yang, et al. For a building structure subjected to an earthquake, the performance of the proposed optimal control law is evaluated and demonstrated by comparing numerically with other available optimal control laws using both deterministic and stochastic earthquake excitations [Refs. 5-7]. Numerical results indicate that the performance of the proposed optimal control law is as good as that of other optimal control laws currently available. However, the contribution of such an optimal control law to the practical implementation of active control systems for seismic hazard mitigations may be quite significant.

Most passive protective systems and devices, such as leadcore elastomeric bearings and sliding systems, are inelastic or non-linear in nature. Hence, the entire structural system, including the hybrid protective system, involves active control of nonlinear structures. Traditionally, control theories are developed for linear systems, whereas control theories for nonlinear systems are very limited. Recently, instantaneous optimal control theories for seismic-excited nonlinear structures have been developed by Yang, et al. [Refs 8-9]. These control theories represent the first step toward the solution of problems associated with hybrid protective systems. Many problems remain to be investigated and applications to various types of nonlinear and inelastic aseismic hybrid protective systems have not been demonstrated.

We have developed two improved optimal control theories for nonlinear and hysteretic hybrid protective systems [Refs. 10-11]. Extensive sensitivity studies are being conducted for applications to different types of hybrid protective systems. Preliminary numerical results indicate that the optimal control theories developed herein are efficient and robust. The research results presented [Refs 10-12] will have a significant contribution to the current state-of-the-art in structural control.

ACKNOWLEDGEMENT

This research is supported by the National Science Foundation Grant No. BCS-89-04524 and NCEER Contract No. NCEER-89-2202.

REFERENCES

1. Yang, J.N., and Wong, D., "On Aseismic Hybrid Control Systems," Proc. International Conference on Structural Safety and Reliability, ICOSSAR' 89, August 1989, pp. 471-478.

2. Yang, J.N., Danielians, A. and Liu, S.C., "Aseismic Hybrid Control System for Building Structures Under Strong Earthquakes," in Intelligent Structures, edited by K.P. Chong, S.C. Liu and J.C. Li, Elsevier Applied Science, 1990, pp. 179-195, Proceedings of The International Workshop on Intelligent Structures, July 23-26, 1990, Taipei, Taiwan.

3. Yang, J.N. and Danielians, A., "Two Hybrid Control Systems For Building Structures Under Strong Earthquakes," NCEER Technical Report, NCEER-90-0015, 1990, SUNY, Buffalo.

4. Yang, J.N., Danielians, A. and Liu, S.C., "Aseismic Hybrid Control Systems for Building Structures," paper accepted for publication in Journal of Engineering Mechanics, ASCE, April 1991.

5. Yang, J.N., Li, Z. and Liu, S.C., "Instantaneous Optimal Control With Acceleration and Velocity Feedback," paper to appear in Proceedings of the 2nd International Conference on Stochastic Structural Dynamics, May 7-9, 1990 Boca Raton, Florida.

6. Yang, J.N. and Li, Z., "Instantaneous Optimal Control With Acceleration and Velocity Feedback," NCEER Technical Report, NCEER-90-0016, 1990, SUNY, Buffalo.

7. Yang, J.N., Li, Z. and Liu, S.C., "Intantaneous Optimal Control With Acceleration and Velocity Feedback," paper accepted for publication in Journal of Probabilistic Engineering Mechanics, 1991.

8. Yang, J.N., Long, F.X., and Wong, D., "Optimal Control of Nonlinear Structures," Journal of Applied Mechanics, ASME, Vol. 55, Dec. 1988, pp. 931-938.

9. Yang, J.N., Long, F.X. and Wong, D., "Optimal Control of Nonlinear Flexible Structures," National Center For Earthquake Engineering Research, Technical Report No. NCEER-TR-88-0002, January 1988.

10. Yang, J.N., Li, Z., and Liu, S.C., "Instantaneous Optimal Control For Nonlinear Hybrid Protective Systems," paper to be presented at the Eighth VPI&SU Symposium on Dynamics and Control of Large Structures to be held on May 6-8, 1991, Blacksburg, VA.

11. Yang, J.N., Li, Z., Danielians, A., and Liu, S.C., "Instantaneous Optimal Control of Hysteretic Structural Systems Under Strong Earthquakes," paper to be presented at the IUTAM Symposium on Nonlinear Stochastic Mechanics to be held in July 1991, in Torino, Italy.

12. Danielians A., "Hybrid Control of Seismic-Excited Building Structures," doctoral dissertation to be submitted to School of Engineering and Applied Science, The George Washington University, in January 1991.

Summary of Presentation at
U.S. National Workshop on Structural Control Research
Los Angeles, California, 25-26 October 1990

CIVIL ENGINEERING APPLICATIONS OF STRUCTURAL CONTROL

by James T. P. Yao, P.E.
Professor of Civil Engineering
Texas A&M University
College Station, TX 77843-3136

1. General Remarks

In recent years, much progress has been made in structural control. Several practical applications either are imminent or have already been made in the analysis, design, and construction of various types of structures including tall buildings, space construction, and ocean structures.

In this presentation, the term "structural control" is used in a broad sense to include the application of active control to civil engineering systems in general. At the risk of repeating existing information and knowledge, several new challenges to the civil engineering profession are outlined and described herein. I plan to discuss several possible applications of structural control in response to these challenges. Moreover, I would like to discuss the effect of the development of structural control on civil engineering education during the 21st Century.

The participants of this workshop and readers of this written summary are invited to explore any of these ideas and to advise me of modern developments. Any discussions and/or insights into such potential applications that I might receive will be appreciated.

2. Civil Engineering Challenges and Applications

There has been much work done to study and develop the intelligent vehicle and highway system (IVHS) all around the world during this past decade. By definition, such a system is the result of applying feedback and automatic control to the highway transportation system. At present, the main purpose of developing this system seems to be the improvement of traffic flow on our highways. As a structural engineer, I envision the enhancement of human safety and/or structural reliability of not only bridges but also guardrails and other components of the transportation system in general. For example, much work has been done on the design of guardrails and other barriers to absorb energy in order to save lives whenever a vehicle accidentally collides with them. It is possible to add active control devices (e.g., air bags) to such barriers to further enhance their lifesaving capability. As another example, it is conceivable to have active devices to protect the highway and/or railroads if imminent landslides can be sensed and their damaging effects can be minimized or mitigated automatically. As yet another (perhaps far-fetched) idea, it might be possible to install active systems in highway pavements to provide automatic repair of potholes whenever they occur (note that there are already tires with the capability to fill punctured holes).

Our society is facing increasing adverse effects of wastes (both solid and liquid) which are potentially hazardous to our health and lives. One of the current challenges to geotechnical and environmental engineers in civil engineering is to design "safe" landfills so undesirable substances will not leach into our groundwater within the design period. It may be possible to provide redundant active systems for such landfills in order to obtain a higher level of safety for the society.

We are all aware of the fact that our infrastructure has been aging and deteriorating for a long time. At present, useful devices such as "smart pigs" have been used by the oil pipeline industry to help detect possible corrosion of steel pipes. Such devices seem to be potentially useful in detecting corrosion and other defects of water pipes and sewer lines as well. Moreover, it seems to be possible to modify and use such devices not only to detect but also to repair various types of pipelines. This concept can also be extended eventually to other components of the infrastructure.

3.   Effects on Civil Engineering Education

To prepare future civil engineers to implement and develop structural control, we need to emphasize and/or add course materials on dynamics, optical and electronic instrumentation, control theory, and decision logic. At present, many engineers are actively discussing and debating civil engineering curriculum and other important matters of our educational programs. While many civil engineers fully realize the importance of improving our curriculum to face the many challenges of the 21st Century, it is not immediately apparent that the desirable changes will be materialized easily in the near future. We must try harder to convey to our fellow civil engineers such a need for changing and modernizing our curriculum. Participants of this workshop are especially well qualified to influence such changes in the civil engineering educational process.

4.   Acknowledgment

I appreciate very much having this opportunity to attend this important workshop. My limited knowledge on structural control results from several research projects supported by NSF during the past two decades and the National Center for Earthquake Engineering Research at SUNY - Buffalo more recently. I have learned a great deal from my colleagues at Texas A&M University since I joined its Civil Engineering Faculty since 1988. Loren Lutes and Norris Stubbs have read this abstract and gave me thoughtful comments and suggestions.

A STUDY ON SYSTEM INDENTIFICATION OF HIGH VOLTAGE CIRCUIT BREAKER AND LIGHTNING ARRESTER WITH ISOLATOR AND THEIR ASEISMIC BEHAVIOR

ZHU BOLONG   XU XIAOYAN
Research Institute of Engineering Structure,
Tongji University, P.R.C.

ABSTRACT

In order to prevent the high voltage (110kv) circuit breaker (CB) and lightning arrester (LA) from breaking down during strong earthquake, two kinds of isolators have been developed. The behavior of the isolators has been got through shaking table test, then dynamic parameters of the two structures have been obtained from system identification and the isolating effects under three kinds of site conditions are predicted.

SHAKING TABLE TEST

Bolt-rubber washer (Fig.1) and moving ball (Fig.2) isolators are used for circuit breaker (CB) at bottom together, and the bolt-rubber washer (Fig.3) isolators are used for lightning arrester (CA) in every connection.
Isolating mechanism of the isolators mentioned above is as follows:
1. By making the structures flexible enough to avoid the predominant period of the soil.
2. By using damping rubber washer to absorb the energy.
3. Through the deformation of the connecting bolts to absorb the energy.

Shaking table test have been undertaken to verify the behavior of the isolators. Following are the test program: (1) the prototype test without installing isolators, (2) the isolating test, which include: (a) skimming with the equal amplitude and changing frequency sine wave, (b) inputting E-C (El-centro 1940 N-S record) as base acceleration.

The comparison of acceleration and strain responses before and after isolation excited by E-C with peak value of 0.3g is shown in Fig.4 and Fig.5. The frequencies before and after isolation are listed in Table 1.

Frequencies Before and After Isolation   Table 1

Equipment	Circuit Breaker	Lightning Arrester
Before Isolation	3.7Hz	5.3Hz
After Isolation	1.0Hz	1.3Hz

The isolation efficiency of the isolators is defined as:

$$\eta = \frac{a_1 - a_2}{a_1} \times 100\%$$

in which a_1 and a_2 are the peak value of response before and after isolation, respectively.
The efficiency of isolators designed in this paper under the excitation of E-C with peak value of 0.3g and 0.4g is shown in Table 2 and 3.

As shown in Table 2 and 3, the isolating efficiency of acceleration (ηa)

on the top of the structures and that of strain (ηa) at the bottom of the
procelatin insulating bushing of the two structures do not agree. This might
be due to the high mode response of the two structures. Hence the reduction
of damaging quantity of structures should be regarded as the standard in judging the isolating efficiency. In this paper the strain (or stress) at the
bottom is chosen as the standard.

Efficiency of Isolator for Circuit Breakser   Table 2

Peak Value of Base Acceleration	Strain (μ)			Acceleration (cm/s^2)		
	Before Isolation	After Isolation	Efficiency ηs	Before Isolation	After Isolation	Efficiency ηa
0.3g	54.29	19.61	64%	765.23	366.45	50%
0.4g	63.09	39.15	38%	979.59	804.75	18%

Efficiency of Isolacor for Lightning Arrester   Table 3

Peak Value of Base Acceleration	Strain (μ)			Acceleration (cm/s^2)		
	Before Isolation	After Isolation	Efficiency ηs	Before Isolation	After Isolation	Efficiency a
0.3g	112.63	36.88	67%	1113.70	753.25	32%
0.4g	166.65	50.78	70%	1381.97	455.07	67%

ANALYSIS

The equation of motion can be shown as follows:

$$[M]\,\{\ddot{X}\} + [C]\,\{\dot{X}\} + [K]\,\{X\} = -[M]\,\{I\}\,\ddot{X}_g$$

Where [M] is the mass matrix; [C] and [K] are the damping and stiffness
matrix of structures; {I} is a unit vector; X_g is the base acceleration; {X},
{$\dot{X}$} and {$\ddot{X}$} are the vectors for relative displacement, velocity and acceleration of mass points, respectively.

A new restoring force model of the connecting point of lightning arrester
and that of sliding and rocking spring of isloater for circuit breaker are
developed.

The following phenomena are found in the test: the unsymmetrical arrangement of flange plate bolts of lightning arrester can make the connecting
point stiffness different in two directions: the static friction between bolt
and installing hole can produce a big initial stiffness, once the friction
disappears, the stiffness will reduce obviously, then a "loss of stiffness"
can be caused, which is not restorable and very helpful to isolation.

A computer program has been developed to identify the isolated system
and to predicte the dynamic response for the isolated equipment, which has
taken into account the phenomena mentioned above.

In the process of system identification a modified POWELL method is used
to identified the dynamic parameters of the structures.

The measured and identified time history of acceleration on the top of
the structures are compared with in Fig.6 and 7.

p 249

CONCLUSION

1. The isolators designed in this paper work well and can reduce seismic action by thirty percent to seventy percent.
2. The efficiency of isolation can be reliably predicted by using the dynamic parameters obtained from the system identification.
3. The efficiency of isolation should be defined as the reduction of damaging quantity of structures with isolators under the same seismic action.

REFERENCE

[1] D.M. Hammelblau "Applied Nonlinear Programming", McGraw-Hill Book Company, 1972
[2] R.W. Clough, J.Penzien "Dynamics of Structures", McGraw-Hill inc. 1975
[3] Zhu BoLong, Xu Xiaoyan "A Study on System Identifcation of High Voltage Circuit Breaker and Lightning Arrester under Seismic Action and Their Aseismic Behavior" Research report of Research Institute of Engineering Structure, Tongji University, 1988

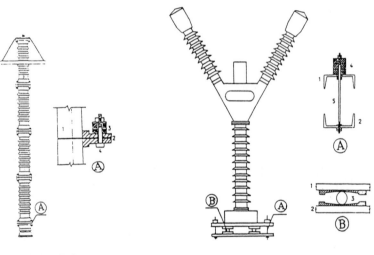

LIGHTNING ARRESTER   CIRCUIT BREAKER

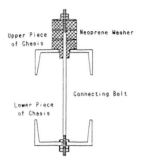

Upper Piece of Chasis   Neoprene Washer

Connecting Bolt

Lower Piece of Chasis

Fig.1 The Connecting of Upper and Lower
Piece of Chasis of Isolator for Circuit Breaker

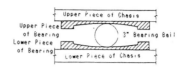

Upper Piece of Chasis
Upper Piece of Bearing
Lower Piece of Bearing   3" Bearing Ball
Lower Piece of Chasis

Fig.2 Bearing Ball and Bearing Piecese

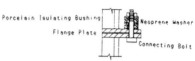

Porcelain Isulating Bushing   Neoprene Washer
Flange Plate   Connecting Bolt

Fig.3 The Connecting of
Flange Plate of Lightning Arrester

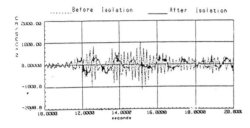

.... Before Isolation _____ After Isolation

Fig.4 The Acceleration on the
Top of Lightning Arrester Excited
by E-C with Peak Value of 0.3g

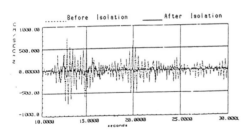

.... Before Isolation _____ After Isolation

Fig.5 The Acceleration on the Top
of Circuit Breaker Excited by E-C with
Peak Value of 0.3g

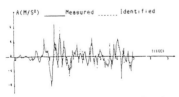

A(M/S²) _____ Measured .... Identified

Fig.6 The Acceleration on the Top of
Circuit Breaker Excited by E-C with Peak Value of 0.1g

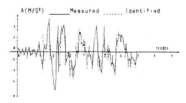

A(M/S²) _____ Measured .... Identified

Fig.7 The Acceleration on the Top of
Lightning Arrester Excited by E-C with Peak Value of 0.4g

p 251

APPENDIX A

Workshop Announcement

U.S. NATIONAL WORKSHOP
ON
STRUCTURAL CONTROL RESEARCH

25 - 26 October 1990
Davidson Conference Center
University of Southern California
Los Angeles, California

OBJECTIVE:

In recognition of the growing awareness by civil engineers worldwide of the potential of active (hybrid) protective systems for natural hazard mitigation, the U.S. National Science Foundation has established a "Panel on Structural Control Research" to (1) develop a plan for a U.S. program in the active control of civil structures and (2) develop a plan under the auspices of the International Decade for Natural Disaster Reduction for U.S. participation in collaborative international research in the active control field.

In order to carry out its responsibilities, the U.S. Panel will convene a Workshop on Structural Control Research with the following major objectives:

1. To summarize the state of the art of the structural control field.

2. To identify and prioritize needed research in the field.

3. To develop preliminary plans for the analytical and experimental advancement of the field and for the performance of full-scale testing.

4. To facilitate the transmission of information concerning the state-of-the-art developments in this rapidly evolving field.

PROGRAM:

The first day of this two-day workshop will feature invited papers and summary reports by U.S. and international representatives from universities, research organizations and industrial concerns. The second day of the conference will consist of several workshop panels that will discuss topics related to specific areas such as (1) analytical methods, (2) experimental approaches, (3) building applications, and (4) non-building applications. Ample time is scheduled for participant questions and interactions. The Workshop will culminate in a detailed, prioritized research plan in addition to recommendations for collaborative research efforts.

PROCEEDINGS:

Post conference proceedings will include invited lectures, summaries of presentations by participants, research needs, and workshop conclusions.

PAPER SUBMITTAL:

Individuals wishing to make a brief presentation on pertinent topics are requested to submit an extended summary of their paper by 1 September 1990. The Workshop

Steering Committee will select for presentation a representative number of the submitted papers. Presentations about ongoing research activities or future research plans in the general area of adaptive structural systems are solicited. Among the specific topics to be addressed in the Workshop are:

1. possible applications

2. reliability issues

3. combined active/passive approaches

4. large scale tests

5. economic issues

6. retrofitting problems

7. implementation techniques

8. control purpose (safety/comfort)

9. dynamic environment (wind, earthquake, etc.)

10. actuator development

11. distributed sensor technology

12. control energy sources

13. control algorithms

14. active parameter control methods

15. stability considerations

16. system identification procedures

17. computational issues

18. architectural considerations

19. educational/training issues

20. international collaborative projects

ATTENDANCE:

The program is designed for researchers, engineers, architects, and others who are interested in the theory, experimentation, and application of the vibrational control of structures, sensitive equipment, etc., under the action of earthquake and wind.

ORGANIZATION:

The workshop is organized by the U.S. Panel on Structural Control Research and supported by the Earthquake Engineering Program of the U.S. National Science Foundation through a grant to California Universities for Earthquake Engineering Research (CUREe). The NSF program director is Dr. S.C. Liu. Arrangements for the Workshop will be handled by the Department of Civil Engineering at the University of Southern California.

The U.S. Panel on Structural Control Research is composed of :

- Mr. William M. Arden (MTS Systems Corporation)
- Professor William Hall (University of Illinois)
- Professor George W. Housner (California Institute of Technology)
- Professor Sami F. Masri (University of Southern California)
- Professor Masanobu Shinozuka (Princeton University)
- Professor Tsu T. Soong (State University of New York at Buffalo)
- Mr. Ben K. Wada (Jet Propulsion Laboratory)

REGISTRATION:

The registration fee for the two–day workshop is $75 if postmarked by 30 September 1990. Registration after this deadline is $100 and subject to space availability. Fees include two luncheons, coffee breaks, proceedings, handouts and attendance at all sessions. An optional evening banquet is planned for which an additional charge of $25 will be made. Payment or purchase order must be received with the registration to guarantee space in the workshop.

For further information, please call:

Mrs. Samia Issa
Workshop Administrator
Telephone: (213) 743-4954
FAX: (213) 744-1426

Or write to:

Control Workshop
Department of Civil Engineering
University of Southern California
Los Angeles, California 90089-2531

Cancellations prior to 3 September 1990 are subject to a $10 handling fee. Refunds will not be made for cancellations after this time.

LODGING:

Block reservations have been made at the University Hilton Hotel adjacent to the USC campus. The rates are $65 single and $75 double. Call (213) 748-4141 for reservations. Payment for lodging must be made directly to the hotel of your choice. A complete listing of other motels/hotels in the area will be included with the registration confirmation letter together with information on transportation and recreational sights in the Los Angeles area.

APPENDIX B

List of Participants

LIST OF PARTICIPANTS

Prof. A. M. Abdel-Ghaffar   University of Southern California
Department of Civil Engineering
Los Angeles, California   90089-2531

Dr. Samy A. Adham   Agbabian Associates
1111 Arroyo Parkway
Pasadena, California   91105

Prof. Mihran S. Agbabian   University of Southern California
Department of Civil Engineering
Los Angeles, California   90089-2531

Dr. A. Emin Aktan   University of Cincinnati
Dept. of Civil and Environmental Eng.
741 Baldwin (ML 71)
Cincinnati, Ohio   45221-0071

Prof. Haluk M. Aktan   Wayne State University
Civil.Engineering Department
Detroit, Michigan   48202

Thomas L. Anderson   Fluor Daniel
Irvine, California

Prof. Jim Beck   California Institute of Technology
Caltech 104-44
Pasadena, California   91125

Lee Benuska   Lindvall Richter Associates
825 Colorado Blvd., Suite 114
Los Angeles, California   90041

Prof. Lawrence A. Bergman   University of Illinois
110 Transportation Building
Department of Aeronautical and

Astronautical Engineering
104 South Mathews Avenue
Urbana, Illinois   61801

Raimondo Betti

University of Southern California
Department of Civil Engineering
Los Angeles, California   90089-2531

Robert Bruce

Applied Technology Council
3 Twin Dolphin Drive, Suite 275
Redwood City, California   94065

Dr. Ian G. Buckle

NCEER
105 Red Jacket Quadrangle
Buffalo, New York   14261

Prof. Thomas K. Caughey

California Institute of Technology
Div. of Eng. & Applied Science
Pasadena, California

Dr. Charles Chassaing

Radix Systems, Inc.
2 Taft Court
Rockville, Maryland   20850

Prof. A. G. Chassiakos

Calif. State Univ. Long Beach
Long Beach, California

Prof. Huei Tsyr Chen

National Central University
Department of Civil Engineering
Chungli, Taiwan   32054
Taiwan, REPUBLIC OF CHINA

Dr. Franklin Y. Cheng

University of Missouri-Rolla
Civil Engineering Department
Rolla, Missouri   65401

Yang Chongzon

CHINA

Dr. J. Edward Colgate

Northwestern University
Department of Mechanical Engineering
2145 Sheriden Road
Evanston, Illinois   60208

Dr. Joel P. Conte

Rice University
Department of Civil Engineering
P. O. Box 1892
Houston, Texas   77251

David J. Dowdell

University of British Columbia
2324 Main Mall
Vancouver, B. C. V6TIW5 CANADA

Abraham Ellstein

Laboratorios TLALLI, S. A. DE C. V.
Planta Temazcal # 10
Tlalnepantla, Mex. 54060 MEXICO

Dr. Akira Endoh

Kajima Corporation
Kobori Research Complex
6-5-30 Akasaka, Minato-ku
Tokyo, 107   JAPAN

Ms. Q. Feng

Princeton University
Department of Civil Engineering and
        Operations Research
Princeton , New Jersey

Prof. Douglas A. Foutch

University of Illinois at Urbana
3129 Newmark Lab.
205 North Mathews Avenue
Urbana, Illinois   61801

Prof. Dan M. Frangopol

University of Colorado at Boulder
Civil Engineering Department
Center OT 4-21

Campus Box 428
Boulder, Colorado   80309-0428

Behram Gonen

GSA, PBS, D & C, 9 PCPE
525 Market Street
San Francisco, California   94105

Dr. Jacob S. Grossman

Robert Rosenwasser Assc. P. C.
1040 Avenue of the Americas
New York City, New York   10018

Mr. Ricardo Guzman

OSA/Structural Safety Section
301 Howard Street, Suite 400
San Francisco, California   94105

Prof. William J. Hall

University of Illinois at Urbana
Department of Civil Engineering
205 North Mathews Avenue
Urbana, Illinois   61801

Prof. Robert Hanson

National Science Foundation
National Science Foundation
Washington D. C.

Prof. Aslaug Haraldsdottir

University of Washington
Department of Mechanical Engineering
FU-10
Seattle, Washington   98195

Prof. G. C. Hart

University of California, Los Angeles
Department of Civil Engineering
Los Angeles, California   90024

John R. Hayes, Jr.

US Army Constr. Eng. Res. Lab.
Att: CECER-EM-E
P. O. Box 4005
Champaign, Illinois   61853

Dr. Alan G. Hernried

James A. Hill

Dipl. -Ing. Gerhard Hirsch

Dr. Zhikun Hou

Prof. George W. Housner

R. Joe Hunt

Prof. H. Iemura

Dr. Yasuo Inada

Oregon State University
Department of Civil Engineering
App. Hall 206
Corvallis, Oregon   97331

James A. Hill & Associates, Inc.
1349 East 28th Street
Los Angeles, California 90806

Aachen University of Technology
Leichtbau
Aachen,   5100   GERMANY

California Insitute of Technology
Department of Civil Engineering
Thomas, 104-44
Pasadena , California   91125

California Institute of Technology
Pasadena, California

Martin Marietta Energy Systems
P. O. Box 2009
Oak Ridge, Tn   37931-8037

Kyoto University
School of Engineering
Kyoto, 606   JAPAN

Shimizu Corporation
Institute of Technology
4-17, Etchujima 3-Chome
Koto-ku
Tokyo, JAPAN

Mr. Jose Antonio Inaudi

University of California, Berkeley
Department of Civil Engineering
National Inf. Service for EQ. Engr.
1301 South 46th Street
Richmond, California   94804

Dr. M. Saiful Islam

Englekirk & Hart, Inc.
2116 Arlington Avenue
Los Angeles, California   90034

Prof. W. D. Iwan

California Institute of Technology
Div. of Eng. & Applied Science
Pasadena, California

Prof. Der-Shin Juang

National Central University
Department of Civil Engineering
Chung-Li,   32054
Taiwan, REPUBLIC OF CHINA

Lambros Katafygiotis

California Institute of Technology
Caltech 104-44
Pasadena, California   91125

Dr. Soichi Kawamura

Taisei Corporation
Earthquake Engineering Team
Technology Research Center
344-1 Nase-machi, Totsuka-ku
Yokohama, 245   JAPAN

Prof. James M. Kelly

University of California, Berkeley
EERI
1301 South 46th Street
Richmond, California   94804

Prof. Anne S. Kiremidjian

Stanford University
Department of Civil Engineering
Terman 238
Stanford, California   94305

Prof. Takuji Kobori

Kajima Corporation
Kobori Research Complex
5-30 , Akasaka 6-chome,
Minato-ku
Tokyo, 107   JAPAN

Nohiride Koshika

Kajima Corporation
Kobory Research Complex
KI Building,
6-5-30 Akasaka, Minato-ku
Tokyo, 107   JAPAN

Prof. John Kosmatka

University of California, San Diego
San Diego, California

Wei Lian

CHINA

Dr. Albert N. Lin

Natnl Instit Standard & Technology
Bldg 226, Room B168
Gaithersburg, Maryland   20899

Dr. S. C. Liu

National Science Foundation
Washington D. C.

Prof. Chin-Hsiung Loh

National Taiwan University
Center for E.Q. Research
Department of Civil Engineering
Taipei, Taiwan, REPUBLIC OF CHINA

Gong Kiu Long

CHINA

Prof. J. E. Luco

University of California, San Diego
Department of Applied Mechanics and
Engineering Sciences
La Jolla, California   92093-0411

David C. Ma

Argonne National Laboratory
9700 S. Cass Avenue
Argonne, Illinois 60439

Prof. Sami F. Masri

University of Southern California
Department of Civil Engineering
University Park
Los Angeles, California 90089-2531

Prof. Leonard Meirovitch

VPI
Dept Engin Science and Mechanics
Blacksburg, Virginia 24061-0219

Dr. Joerg Melcher

German Aerospace Res. Establishment
(DLR), Bunsenstr. 10
Goettingen, 3400
GERMANY

Prof. Richard K. Miller

University of Southern California
Department of Civil Engineering
Los Angeles, California 90089-2531

Dr. Akira Mita

Shimizu Corporation
Ohsaki Research Institute
Fukoku Selmel bldg. 2-2-2
Uchisaiwai-cho, Chiyoda-ku
Tokyo, 100 JAPAN

Prof. Shin Morishita

Princeton University
Department of Civil Engineering and
        Operations Research
School of Engineering and Applied Science
Princeton, New Jersey 08544

Koji Naraoka

Shimizu Corporation
Ohsaki Research Institute

2-2-2 Uchisaiwai-cho
Chiyoda-ku
Tokyo, 100   JAPAN

Prof. David Nemir

The University of Texas at El Paso
500 W. University Avenue
Electrical Engineering Department
El Paso, Texas   79968-0523

Dr. Robert Nigbor

Agbabian Associates
1111 S. Arroyo Pkwy, Suite 405
Pasadena, California   91105

Yoichi Nojiri

Kajima Institute of Construction Tech
Kajima Corporation
19-1 Tobitakyu 2-Chome, Chofu-Shi
Tokyo, 182   JAPAN

Prof. Irving J. Oppenheim

Carnegie Institute of Tecnology
Department of Civil Engineering
Carnegie Mellon University
Pittsburgh, Pennsylvania   15213-3890

Prof. Benito M. Pacheco

University of Tokyo
Department of Civil Engineering
7-3-1 Hongo, Bunkyo-ku
Tokyo, 113   JAPAN

Dr. A. S. Papageorgiou

Rensselaer Polytechnic Institute
Department of Civil Engineering
JEC 4049
110, 8th Street
Troy, New York   12180-3590

Victor M. Pavon

Consultores del Concreto
P. O. Box 74-A
Satelite,   53100
MEXICO

Telephone Number: 905-572-5571

Dr. Zolan Prucz

Modjeski and Masters
Lousiana

Prof. Jun Ping Pu

Feng Chia University
Department of Civil Engineering
University of California, Berkeley,
1301 South 46th Street
Richmond, California 94804-4698

Dr. Jay J. Pulli

Radix Systems, Inc.
2 Taft Court
Rockville, Maryland 20850

S. Z. Qi

CHINA

Prof. A. Reinhorn

State University of New York, Buffalo
Department of Civil Engineering and
      Applied Sciences
212 Ketter Hall
Buffalo, New York 14260

Cesar Alfonso Reyes Zamora

CICESE / Department of Seismology
Espinoza 843
Ensenada, B. C.
MEXICO

Jim Rides

Unicersity of California at San Diego
Department of AMES - MC:R011
La Jolla, California 92093

Prof. Jose Rodellar

Tech Univ of Catalunya
Barcelona,
SPAIN

Ms. Philippa Rogers

Science and Engin. Research Council
Washington, D.C.

Prof. David H. Sanders

University of Nevada
Civil Engineering Department
Reno, Nevada   89557

Prof. Robert H. Scanlan

Johns Hopkins University
Department of Civil Engineering
Baltimore, Maryland

Prof. Nisar Shaikh

University of Nebraska
Department of Civil Engineering
212 Bancroft Bldg.
Lincoln, Nebraska   68588-0347

Li-Hong Sheng

Department of Transportation
P. O. Box 942874
Sacramento, California   94274-001

Dr. Keiji Shiba

Shimizu Corporation
Institute of Technology
4-17, Etchujima 3-Chome
Koto-ku
Tokyo, JAPAN

Dr. Benson Shing

University of Colorado
Department of Civil Engineering
Campus Box 428
Boulder, Colorado   80309

Prof. Masanobu Shinozuka

Princeton University
Department of Civil Engineering and
    Operations Research
School of Engin and Applied Science
Princeton, New Jersey   08544

Chen Shouliang

CHINA

Prof. H. Allison Smith

Stanford University
Department of Civil Engineering
Terman Engineering Center
Stanford, California 94305-4020

Prof. Tsu T. Soong

State University of New York, Buffalo
Buffalo, New York 14260

Prof. B. F. Spencer

University of Notre Dame
Deaprtment of Civil Engineering
Notre Dame, Indiana 46556

Prof. John F. Stanton

University of Washington
Department of Civil Engineering
233 More Hall, FX-10
Seattle , Washington 98195

Masaharu Sugano

Hitachi, Ltd
Tsuchiura Works,
603, Kandatsu-machi
Tsuchiura-shi, Ibaraki-ken, 300 JAPAN

Dr. Tetsuo Suzuki

Obayashi Corporation
Technical Research Institute
4-640, Shimokiyoto
Kiyose-shi
Tokyo, 204 JAPAN

Prof. Costas Synolakis

University of Southern California
Department of Civil Engineering
Los Angeles, California 90089-2531

Prof. Iradj G. Tadjbakhsh

Reinsselaer Polytechnic Institute
8th Street
Troy, New York 12181

Yoshitaka Takeuchi

Stanford University
Civil Engineering Department
Quillen # 7A
Escondido Village
Stanford, California   94305

Albert Tung

Department of Civil Engineering
San Jose State University
One Washington Square
San Jose, California   95192-0083

Mr. John Vostrez

Office of New Technology
Division of New Technology
Caltrans
5900 Folsom Blvd.
Sacramento, California   95819

Ben Wada

Jet Propulsion Lab.
Pasadena, California

Arthur Wong

City of Los Angeles
Department of Building & Safety
Rm 422 City Hall
200 N. Spring Street
Los Angeles, California   90012

Prof. H. L. Wong

University of Southern California
Department of Civil Engineering
Los Angeles, California   90089-2531

Dr. Hiroyuki Yamanouchi

Building Research Institute
Ministry of Construction
1 Tatehara
Tsukuba-city
Ibaraki, 305   JAPAN
Telephone Number: 0298-64-2151
Fax Number: 298-64-2989

Prof. C. Y. Yang

University of Delaware
Civil Engineering Department
Newark, Delaware   19716

Prof. J. N. Yang

The George Washington University
Department of Civil Engineering
Washington, D. C.   20052

Prof. J. T. P. Yao

Texas A. &M. University
Department of Civil Engineering
College Station, Texas   77843-3136

Mr. Mark Yashinsky

Department of Transportation
P. O. Box 942874
Sacramento, California   94274-0001

Nabih F. Youssef

Nabih Youssef & Associates
660 S. Figueroa
Los Angeles, California

P. L. Zou

CHINA